The Gardens of
Arne Maynard

The Gardens of Arne Maynard

Arne Maynard

Photography by William Collinson

Foreword by Rosie Atkins

MERRELL
LONDON • NEW YORK

CONTENTS

I dedicate this book to my dear friend David Abbott, who gave me the opportunity to create my first garden for him and his wife, Eve, at Hunworth Mill in Norfolk. His faith in me set me on this wonderful journey through gardens.

ARNE MAYNARD

Foreword

I first met Arne Maynard in 1991 while I was submerged in the exhilarating process of launching *Gardens Illustrated* magazine. Arne was just starting out as a garden designer, and I loved the way he enthused about his horticultural heroes, William Kent, Harold Peto, Lady Salisbury and Lady Arabella Lennox-Boyd, the last of whom he credits with turning garden design into a profession.

Around this time Arne and his partner, William Collinson, bought an almost derelict sixteenth-century manor house in the fens. Guanock House was a garden-magazine editor's dream, and William's pictures, taken for this book, instantly transport me to the house and garden the pair brought back to life.

The late David Abbott, to whom this book is dedicated, and his wife, Eve, were early clients, and Arne designed a wonderful garden around their eighteenth-century corn mill in Norfolk. Nearly thirty years on it is possible to spot recurring motifs in Arne's gardens: the handcrafted oak gates and fence posts, bespoke greenhouses, clipped trees and timeless planting.

Arne has always enjoyed developing a close relationship with his clients and their gardeners, even when the gardens are open to the public, as is Haddon Hall in Derbyshire. This garden beautifully enhances one of Britain's most romantic historic houses. By contrast, and demonstrating Arne's versatility, La Frateria in Italy reflects the owner's taste for the contemporary. The striking arbour is a shady place to eat and contemplate the view, and the planted terraces and modern greenhouse harmonize with the landscape.

In 2000 Arne and the Dutch garden designer Piet Oudolf collaborated on a garden for *Gardens Illustrated* at Chelsea Flower Show, winning a Gold Medal and the award for Best in Show. Their garden showcased cloud hedging, which Arne uses to dramatic effect in his garden for a Queen Anne house in West Sussex. Arne went on to win another Gold Medal at Chelsea, and he is now one of the judges.

In 2006 Arne and William left the fens to create a new garden in the Welsh hills. Allt-y-bela, their medieval tower house, has become a place of pilgrimage for new gardeners and head gardeners seeking inspiration or wanting to hone their gardening skills. You can stay overnight when there are no workshops in progress, and I am sure it won't be long before the website announces plays to be performed in the new outdoor theatre.

Arne and his loyal and talented team continue to attract clients from around the world. This book is a fitting record of how this highly acclaimed designer has refined his skill over the years while celebrating his joy in sharing his ideas with others.

ROSIE ATKINS

Roses tumbling over the building create a relaxed, informal atmosphere at Hunworth Mill in Norfolk.

Introduction

My very earliest memories are of gardening. I remember holding a basket and walking behind my grandfather as he weeded his vegetable garden, and being distraught that he was pulling out carrots to put on the compost heap. I tried to plant them out again, drowning them in water, hoping that they would grow. Of course they did not, but, undeterred, I spent my whole childhood gardening.

For me it was a welcome escape from sport, and by the age of about seven I had the job of looking after the school garden. I earned all my pocket money from gardening and mowing lawns, usually working alongside the elderly people I was helping. Many had gardened all their lives, and were keen to pass on their knowledge. One lady showed me how her tulips were exactly the colour of the pink in the apple-blossom bud in the tree above. That was the very start of my awareness of how to look at the details in gardens. By the age of nine or ten I was starting to choose plants for myself, and I strapped baskets and boxes to my bike so that I could carry back plants from nurseries and from the roadside stalls that were once a frequent sight in the country, always offering cabbage plants and wallflowers.

Gardening was already my greatest interest, and I was hungry to know more, so my godmother started taking me to local gardens. The most influential of these visits was to Cranborne Manor, the garden that awoke my passion for garden design; what I learned there has stayed with me. This wonderful old hunting lodge nestles in the Dorset hills, close to the ancient Cranborne Chase, a royal hunting ground from at least the time of William the Conqueror until the seventeenth century. A thrill ran through me the first time I saw the view through the main gate, down the magnificent avenue of mature beech trees that leads to the house. I was bowled over by the beauty and atmosphere.

The seventeenth-century garden was laid out by Mounten Jennings and John Tradescant, the latter of whom supplied many of the original plants. The historic structure of the garden is remarkable, but what I found even more interesting was the way it had been interpreted and gardened by Lady Salisbury, whose clever use of plants made it elegant and relaxed at the same time. Every week I would cycle over and wander around the garden, helping out at the nursery in exchange for a plant or two.

Topiary is one of the essential planted building blocks that add structure and scale to the garden at Allt-y-bela.

PAGES 8–9
In the garden I designed for Laurent-Perrier at the Chelsea Flower Show in 2012, a sculpture by Breon O'Casey was framed by a pleached copper beech walk.

RIGHT AND OPPOSITE
The wonderful combination of formality and informality that runs through the garden at Cranborne Manor in Dorset has created one of the most romantic gardens in England.

I learned about old-fashioned roses and herbaceous perennials, and through familiarity with the place, I started to absorb things like the effect of the pattern-making in the knot garden, and to understand the importance of having areas with different characters. I loved the white garden, where snowy dianthus flowered under the ancient espaliered apples, filling the air with scent. The sundial garden with its raised mound was filled with Hidcote lavender, its outer beds full of roses and hardy geraniums. Next to it were the bowling green and croquet lawn, and adjacent to that the orchard, where I loved the contrast of wild cow parsley, growing in the long grass under the trees, with perfectly clipped yew hedges. The garden was full of variety and drama, and, more than anything else, at Cranborne I learned the importance of creating or emphasizing atmosphere.

Although Lady Salisbury is no longer at Cranborne Manor, and the garden has changed since my childhood, it is still a beautiful place to visit. I feel that this was where I began training my 'eye', and it is still my yardstick for quality and good design. Before this awakening of a certain taste, I had been buying African marigolds and petunias and bedding them out in the manner of a parks department. Now I felt I'd found a new way of gardening, and I started to cycle or drive with my godmother to other gardens in the National Gardens Scheme's *Yellow Book*, burning with enthusiasm to find new plants and learn about different horticultural techniques. My early passion was definitely for horticulture over design, and this deep love of the act of gardening and the plants themselves has never abated. I am more thrilled than ever to discover a new plant, and when I visit gardens or nurseries I always seek out the gardeners and nurserymen and women, because I love to hear at first hand about their knowledge and experiences of different plants.

All this garden-visiting also helped me to recognize the importance of putting the right 'bones' into a garden, and that those structural elements, if planted, must be absolutely the right plant in the right place, for longevity. A good framework – the way you move around the garden, the green architecture and how it relates to the house and the views – will withstand the changes of fashion

RIGHT AND OPPOSITE
A visit to Rousham at any time of year never fails to demonstrate the remarkable timeless quality of William Kent's design.

and moods of the owners. The structure of the garden has to 'hold hands' with the environment, respecting the architecture of the house and the setting, the climate and the culture, whether urban or in the country. To achieve this you must be an observer, and note the detail: to see how plants grow in the local conditions, and how they have been traditionally managed in different regions. This is the information I seek every time I look at a new project, in order to design a garden that will feel right in its setting, and will last.

When I started my garden in Wales, for example, I had to adjust my thinking dramatically, having gardened previously in the flat, dry conditions of East Anglia. It was not only the list of plants themselves that had to be adjusted to suit the wet, hilly landscape of my new home, but also the way those plants were grown. At Guanock House the path to my front door was lined with lavender. It was a magical part of the garden, one that affected all the senses. The lavender was planted so that it almost covered the path, and you had to brush the flowers as you passed, which released great wafts of scent. The colour was, for me, the true indication of summer. I loved the ritual of delaying the flowering by cutting the plants back in early spring to give fresher-looking flowers later in the season, observing the results every year and learning how to tweak the process in order to make the plants perform that little bit better.

It had been a vital part of my gardening year, and I was extremely reluctant to give it up, but I could not impose a lavender-flanked path on Allt-y-bela. Not only would the lavender sulk and eventually rot in the wet clay soil, but also the symmetry of a double row would be at odds with the asymmetry of the building and the plot. I am not completely without lavender at Allt-y-bela: a potted plant seeded itself into my cobbles, and it survives in the drier growing conditions they provide. But this is the only place it feels right. That is what I mean by starting every design with observation, to learn what is right for the setting.

Observing the local conditions, vernacular architecture and plants is one part of beginning a design. The other is the library of images and influences I have built up over many years, and which I carry around in my head. I do photograph them, too, but I am lucky enough to have a strongly visual memory, and when the situation seems right I might recall a detail from twenty years before. To make a new design I reinterpret ideas and put them together in new ways, and so I am always seeking fresh inspiration, and visit new gardens as often as possible. I will always come away having gained something, an insight into a different way of doing things: using plants in a different combination, for example, or laying paving in a different style.

I also take inspiration from old architectural engravings and paintings. I spend a lot of time looking at books, and sometimes a drawing can prompt a more vivid image than a photograph or even seeing something in the flesh. A depiction of a knot garden, picket fence or beautiful gate in an early English portrait of a house might spark an idea, but I must imagine the details. Seeing a pair of gates leading to another (and gates stand out, because they are often magnified to look more impressive in paintings) might start a train of thought that comes out in a design for a client. In my mind I can develop the picture, imagining how the space between the gates should be, how the texture of the latch should feel to the hand and how the paving should look. The sources I have used include the famous engravings by Johannes Kip and Leonard Knyff and the sixteenth-century gardener's labyrinth by Thomas Hill, the latter of which contains not only visual references to many types of garden, but also fascinating descriptions of gardening techniques, which often trigger ideas.

I also look at paintings, fabrics and sculpture, since they are a rich source of ideas about colour, texture and pattern-making. Feeding my imagination in this way enables me to use images as the starting point for selecting plants when I find the right situation. Sometimes these details are found in the house of the garden on which I'm working, and it is very satisfying to make the link with the interior, or with an object the client has collected.

There is a handful of gardens to which I return steadfastly time and again, that always enrich my thinking and are real favourites. After Cranborne came Rousham House and its garden, which I got to know when I was in Oxford, studying architecture at the polytechnic. I fell in love with the elegant simplicity of the design of this early

eighteenth-century classic of the early landscape movement, and see it as proof that the fundamentals of good design do not change. From my very first visit I was struck by the amazing sense of arrival. The parkland is a pure eighteenth-century idyll, with grazing longhorn cattle, a rusticated temple and a glimpse of a classical doorway in the boundary wall. Past the unpretentious lodge, the view suddenly opens up to the beautiful Jacobean facade of the house, framed by an avenue of huge, mature lime trees. A simple broad swathe of velvety striped grass in front completes the archetypal English country-house scene.

Although Rousham is a big, grand garden, it is all about the strength and simplicity of the design ideas, and I find these principles can be reinterpreted for any garden, on any scale. Only a few species are used to create impact, and that has taught me that a limited palette of plants can produce a very powerful effect. Two or three species of tree, or ten varieties of herbaceous plant in a border, can be more effective and elegant than everything in the nursery catalogue.

Rousham has many superb sculptural pieces, and they are carefully placed to direct movement through the garden, creating an invisible thread that pulls the visitor along. There are no signs in the garden, yet everyone follows the same route. Subtle contrasts are drawn out between light and shadow, sunshine and shade, cut grass and uncut grass, nature and the manmade. Grass is very important in this garden. It is the unifying surface throughout, but is used in many different ways: a mown edge to a border; a simple chamfered bank around a rectangular lawn; long grass alongside short grass; a very narrow grass path that slows you down; a big, wide one that moves you on quickly.

Rousham is not a flowery garden: there are flowers, but they are separate from the rest of the garden. The structure is composed by trees, shrubs and grass, and is therefore present all year round. This is an excellent demonstration of the use of structure in a garden. The rill of cut natural stone, for example, holds a sinuous ribbon of water that snakes in and out of the light, travelling through the dappled shade of overhanging limes and between banks of beautifully clipped Portuguese laurel. The effect of progressing along a shadowy green

OPPOSITE AND ABOVE
The garden at Rousham is created with a restrained palette of plants that bring a great sense of antiquity in their maturity.

tunnel towards a brightly lit object is intensified by the simple palette of plants and materials, and this characteristically eighteenth-century way of using laurel as a structural element is very pure. The laurel creates weight, acting as an under-storey and providing a further contrast with the trees growing wild and free above.

Rousham is a garden that has been very precisely thought out, and the restricted palette of the landscape, its temples and statues, is very closely regulated. In the walled garden, however – despite its architecture and organization – informal planting takes on the character of the people who live in the house, bringing the whole garden alive. This compartmentalization of two distinct garden types within one garden is something else I've taken from Rousham. The formality of a restricted palette of plants in one area of the garden can be beautifully complemented with a separate garden, behind walls or hedges, where the planting can be as wild, loose and free as you like. The transition is rather like crossing a moat, and the colours of the herbaceous border are very striking after all the green. But in fact they are quite subtle, what I'd describe as faded-chintz colours, with roses, *Alchemilla mollis*, campanulas, asters and other simple, unshowy plants. It is not until the third of the walled gardens that the colours really heat up. Growing late summer flowers like asters for cutting in kitchen-garden beds and borders is a country-house tradition, and for me that is where they belong, part of the harvest festival scene, with pumpkins and onions. Even after the flowers have faded, their seed heads give structure to the gently collapsing border and bring the birds in.

For me, the mixture of utility and decoration is the essence of the kitchen garden, and what I love most about Rousham is how it manages this progression from a landscape of grass, water and stone to something that is vibrant and yet still in keeping with the design and purpose of the garden as a whole.

The manner in which Rousham has been maintained – the care and skill of the gardeners – is a key part of the garden as we experience it now. For me this visible love for the art of gardening is what sets really great gardens apart. For a garden to thrive it requires consistent, dedicated maintenance, as well as a steady stream of fresh creativity and a real

RIGHT AND OPPOSITE
Visiting Llowes Court, my friend's garden in Powys, inspires me to look at the many different ways magic can be created in a garden.

sense of enjoyment in the activity of gardening itself. Without these ingredients, however good the design, a garden will always lack vitality. I notice a buzz in gardens where there is an appetite for seeking out unusual plants and shrubs, for constantly refreshing plantings and experimenting with ways to clip, train or support those plants. Gardens that have evolved through this kind of love of gardening, rather than by design, are made by people who cherish their gardens and the environment in which they sit, and are often those that inspire me the most.

I have quite recently come to know a private garden of this kind, and I count it as one of my favourites: Llowes Court, near Hay-on-Wye in Powys. This playful garden, at once restrained and brimming with exuberant romance, is full of art. Verbascums and *Alchemilla mollis* grow out of paving, forget-me-nots seed themselves under apple trees, and herbaceous plants tumble over the edges of borders, but all this contrasts with the formally clipped yews against the grand stone walls, the geometric knot of clipped box, and the topiary by the entrances. This contrasting combination – hard architecture and structural planting (clipped or trained), against the softness of plants that have a relaxed habit and are allowed to behave in a natural way – is something I try very hard to achieve in my own gardens.

It is evident at Llowes Court that this style has evolved naturally. It is one of those gardens I mentioned earlier, in which the design has never been drawn out, but has come about through the act not just of gardening, but also of living in the garden. For instance, when a vegetable garden was needed, it was placed close to the house for practicality, and next to a large yew hedge. But although this gave it some sense of its own space, a hedge on one side did not enclose it enough. Carved oak posts were put in, therefore, and sections of beautiful oak trellis set between them, giving just the right sense of enclosure, with the added benefit of becoming a structure for more roses. More oak posts were then used to hang the washing lines across into the laundry yard next to the vegetable garden. As the garden developed further, a little oak bridge was made to cross the stream, and its detailing repeated that on the posts in the laundry yard and vegetable garden. That detail has now become a signature for the garden, although it was never planned that way, and continues to evolve.

The garden at Llowes Court reflects the taste and passions of the owners, and there is a great sense of not being too pompous or too smart. There are flights of fancy, including a tree house beautifully crafted out of twiggy birch and clinging like a giant bird's nest to the sturdy trunk of a poplar at the edge of a plantation. A refined limewashed pavilion with a beautiful gypsum crystal ceiling is perched high on the walls of the rose garden, and reached via a giant's staircase of rough-hewn stone slabs, giving a small sense of mountaineering achievement with every step. An exquisite shell-lined grotto hidden temptingly within the rocky base of the hillside brings out the child in you as it narrows to a tiny passage through which only the brave can squeeze.

The garden is allowed to be slightly rough at the edges, even a little weedy. Wild flowers are encouraged to seed around, so one might find ox-eye daisies coming up through the roses. I try to breathe this relaxed feeling and lack of preciousness into my gardens. I look for it particularly in country gardens, but even a really sharp city garden feels more genuine if the plants are allowed to have character: if trees with slightly crooked trunks are chosen, or if topiary is clipped into organic rather than geometric shapes.

I think a garden is like a stage set. To put on a play you need a stage, backdrops, props, costumes, players and of course the play itself. Only when all these ingredients come together can the play be performed. The first step is to understand the nature of the play, and that is my most important role as a designer: to interpret the wishes of my clients and understand any influences on the design, such as the landscape and the architecture. Then I can bring my design ideas to the stage.

From that moment the creation of a garden becomes an exciting collaboration with a wide range of people, from landscapers and stonemasons to nurseries and craftsmen, and – most important of all – the gardeners who coax it through the early days and ultimately enable the plants to give their best performance. I enjoy greatly the process of working with craftspeople on every detail of a garden. It might be as simple as a door latch or a hinge on a garden gate, but I'm excited to get it right, and to

produce something unique to the garden. Old gates or benches often bear the mark of the maker, and feel a world apart from the dull 'standard' items that make up the vast majority of what's on offer today. I find it very satisfying to come across a new, beautifully crafted gate with a lovely latch, balanced to close with a satisfying click, with a small detail on the hinge that echoes an architectural detail in the house. Llowes Court has an iron gate in the form of a three-dimensional fig tree. It branches out from the wall, just as a real fig would, and part of it is hinged to swing open. It's both sculpture and gate, and brings a specific identity to that space in the garden.

Finding craftsmen to interpret my designs is not always easy, but when we discover the right person the most exciting aspect is that they bring their own knowledge and flair to the design and produce something that exceeds our expectations. There must be a close pairing of craftsman and task. For example, a wall must not just be beautifully and soundly built; it has to work within the garden, too. We can look at photographs and drawings, and specify what we want, or even visit a local example to analyse the style, but the subtle difference between one waller and another will always affect the outcome. A true craftsman constantly makes decisions as he or she works – how smooth the face should be, how tight the gaps, how regular the size of stone – and, like handwriting, the result will show his or her character.

At Allt-y-bela I was delighted with the new stone walls of the stream and garden theatre. They fit so well that it is as though they have been uncovered, rather than built. At my show garden for Laurent-Perrier at the Chelsea Flower Show in 2012, the walls were made of a sophisticated, smooth Parisian limestone, in regular blocks, with fine mortared gaps in exactly the colour I specified. Both styles of wall were satisfying and right for the situation, but each was different and required different skills.

Walls and gates are part of the architecture of the garden – the backdrops, if you will – but the 'props', the furniture and pots, lights and sculptures, have a more immediate impact. They are sometimes seen as ephemeral objects that come and go with different fashions or owners, but that is not necessarily the case, and I prefer to think of them as an essential part of the design. At Rousham, for example, the

sculptures and benches have remained an integral part of the design, unchanged for centuries.

I enjoy working with craftsmen who are familiar with traditional techniques and even patterns, but bring their own flair and dexterity to the work. Terracotta garden pots, for example, once made by village potters up and down the country, were functional and largely similar, but each had a variation that was particular to the maker. Many had a motif or decoration that was specifically for a local house. Although these village potters have largely disappeared and their pots have been replaced by cheap mould-made imports, there is still a handful of potteries making hand-thrown terracotta in Britain. For Allt-y-bela I found a potter who makes an open basket style of pot with latticework decoration, which fits beautifully with the windows of the house and the lattices that are repeated in paths and hedges throughout the garden. In contrast, I also use huge contemporary pots made by a Belgian potter to give weight, scale and modernity to the garden.

Of all the elements that make up a garden, for me the most important is the plants. I am passionate about going to small, specialist nurseries to find unusual plants, or to search out a better strain of a cultivar. It is worth making the effort to seek out a particularly beautiful plant, even if it flowers for only three or four days of the year. It is like owning a stunning jewel that's worn only once a year: that occasion is always memorable. These rarities give an air of excitement and refinement.

Those of us gardening in Britain are very lucky to have nurserymen and women who are dedicated to growing a very wide range of perennials and roses that allow us to bring great subtlety to the planting palettes of our borders. Then there are the specialist growers: of ferns and hellebores, of orchids and auriculas, of hardy geraniums and pelargoniums – in fact of almost every genus of plant. These specialists, plant collectors and breeders are constantly looking for new species and cultivars to offer us. I think it's important to recognize what a marvellous resource these specialist nurseries are, and appreciate the vital part they play in the quality of our gardens. When I am creating gardens abroad, I often find that the choice of plants is much more limited, and have to create impact with restrained, simple plant selections. It certainly makes me appreciate

OPPOSITE AND ABOVE
I seek out skilled craftsmen and women to make the different elements that we need in our gardens, from creating hand-thrown terracotta pots to sculpting the land and building drystone walls or making finely crafted gates.

RIGHT AND BELOW
Specialist nurseries that grow unusual and rare plants provide us with a diversity and quality of plant material that enables us to make each garden individual and special. Shown here are the snowdrop *Galanthus* 'Ecusson d'Or' and the peony known as Molly the Witch (*Paeonia mlokosewitschii*).

the collaborative relationships I have formed with nurseries at home, and I miss the subtle complexity that is made possible by the sheer range of plants they grow.

Excepting the relationship I have with my clients, the most important collaboration of all during the creation of a new garden has to be with the gardener. Just as every stonemason or potter has an individual style that comes through in his or her work, so the gardener will leave their fingerprint in the way they tend the garden. The planted elements are not static; they change daily, and every day the gardener also makes decisions: to stake a plant in a particular way, to deadhead, cut down, prune, sow new seed, replace a failing plant . . . the list is endless and constant. Because of where my own passions lie, I design gardens that are intended to *be* gardened. They are definitely not low-maintenance. The gardener will and should make their own mark on the garden, and make decisions that will carry it forward. Whenever possible I stay in close contact with the gardeners, and I love to continue to be involved with a garden long after the design has been implemented. But every garden will be shaped by the passion and dedication of the person looking after it every day, and I owe a huge debt of gratitude to my clients and their gardeners who have recognized this and not only fulfil my hopes for the design but also often surpass them.

This is a good opportunity to thank all the people who have worked with dedication and great commitment to create the gardens you see here: everyone in my office and on site, as well as the nurseries, craftsmen and gardeners. It is the magic of their contribution that has made these gardens special. I am eternally grateful to my clients, who have enabled me to pursue my passion for making gardens and to carry on learning and developing what I do, and who have allowed me to share their gardens with you in this book.

OPPOSITE
Particularly when it comes to larger specimen trees and topiary, I find it essential to visit the nurseries and choose the individual plants in person. The shape and habit of the trees have a great impact on the character of the garden, and cannot be portrayed truly in a photograph.

Allt-y-bela

PREVIOUS PAGES
We chose the striking orange limewash to give the house the richness of an Elizabethan jewel. The transformation was astonishing. The house had been a dull white-beige, which jumped out of the landscape and felt very cold. Now it relates to the countryside in an extraordinary way, emanating great warmth and energy.

In 2005 we moved from Guanock House, in the vast, flat, dry landscape of the Fens (see page 254), to Allt-y-bela in the rolling hills of wet Wales. The garden setting could not be more of a contrast. The landscape around Allt-y-bela is imbued with a feeling of ancient husbandry, its hedges and ditches and droves a biography of man working with nature; in East Anglia, on the other hand, the landscape is dominated by farming on an industrial scale. In Wales, ancient oaks and field maples mark the boundaries, and wild orchids are scattered abundantly. The landscape is green, not brown. I had found great beauty in the swaying fields of crops and the vast open skies of the Fens, but by August the fields were ploughed and bare. Living there, what I craved was a landscape more akin to the countryside of my childhood: the meadows, bluebell woods and grazing sheep of Dorset.

The house at Allt-y-bela has no absolutely straight lines, and no symmetry. It has evolved over time. Originally a medieval cruck-frame farmhouse, it became rather grander when the extraordinary tower was added in the early seventeenth century. When we moved in, restoration work had been done to prevent the house from falling down, but the garden was an utterly bare canvas. As I began to design, I found it very difficult to edit my ideas; indeed, it is always more difficult to design your own garden than to make one for a client. For this site, I knew I must be disciplined and distil the vast number of things I love, so I began with a list of my favourites: topiary, roses, fruit trees, wild flowers and kitchen gardens. These were all elements I would include, but above all I needed to be true to the identity of the place. This was not a village house nor a manor house in the Cotswolds; it was a farm.

The natural folds of the landscape roll down to the house and to the stream that winds through the base of the valley. The site had never been properly levelled, and the house would never have had a big formal garden, but would have contained the necessary elements for subsistence: a vegetable garden and an orchard, nut trees for winter foraging and herbs for medicine, cooking and strewing. However, the tower was undoubtedly built for show, a fact that gave me licence to add the formality of clipped topiary and the intricacy of flowers. To one side of the property runs the medieval road to

OPPOSITE, LEFT AND BELOW
In winter, particularly, the earthworks of the natural theatre and canalized stream look graphic and modern; at the same time, however, the contours of the banks resemble the lines in fields where the sheep have trodden their meandering tracks. Because the garden is at the bottom of a valley, heavy rain causes the water to rise very quickly, with such force that it can wash the banks away. All along the stream we found the remains of walls that had protected the banks. As we rebuilt them, the stream's sinuous flow gained architectural merit and the prominence it deserves.

OVERLEAF
In winter topiary plays a crucial role. Sculptural in their own right, these trees also contain and divide the spaces between them, as well as bringing essential weight to the garden to balance the scale of the house.

On the earthworks at the back of the house the spring meadow is full of *Crocus* 'Prins Claus' and *C. chrysanthus* 'Cream Beauty' (above), and tiny reticulate irises (opposite): the purple *Iris* 'George' and *I. reticulata*.

Chepstow. These days it is just a bridleway, but for centuries it would have been a busy route, from where many people would have seen the house. I wanted to open up and celebrate these views of the house once more, but also to connect the garden both physically and visually to the landscape around it.

We immediately removed the walls and estate railings that had been put up in order to sell the house. Such delineation of the plot seemed entirely wrong to me: the heavy boundaries weighed the house down and attempted to shut out the countryside. I wanted the house to breathe, and its garden to be part of the landscape. I introduced fences that were as agricultural and transparent as possible, and strengthened the link to the surrounding land by echoing its character in the garden. The asymmetric plot sits among fields etched with ancient drovers' roads that enabled farmers to move carts of hay or bring their cattle or sheep down from the hills. Further crisscrossing lines over the fields are formed by hedges and animal tracks, patterning the land with an irregular lattice.

Once I felt I understood the important characteristics of the landscape, I began setting down the foundation blocks of the garden. I took the droveway that cut across the hill as the inspiration for new earthworks behind the house, and I used the language of the lattice in plaits of box hedging on both sides of the house (they provide the containment I need for slightly more formal flower beds). I repeated the crisscrossing in cobbled paths through the fruit and flowers in the area I call the cottage garden.

The most important foundation block, however, is topiary. I have used it in place of walls, to create circulation within the garden and to divide in a transparent way. The scale and strength of the topiary helps to anchor the house and to reduce the apparent height of the tower. Close to the house are more formal yews and beech, which hold on to their golden leaves in winter. As the garden dissolves into the landscape I have clipped native hawthorn into topiary, and also hedges, which are cut to reflect the treeline on the brow of the hills beyond.

Other mixed hedges have been re-laid in the traditional manner, over the framework of stumps that are a sign of the work of hedge-layers over the years. On the woodland side, the boundary is formed by a tunnel of very old hazels planted either side

Under the low branches of the hazels come first the snowdrops and then the native narcissi, carpeting the common as winter gives way to spring. I have added cultivated hellebores at the edge of the woodland, keeping them sparsely planted so that they are a real treat to discover.

of the medieval cobbled road, forming a wonderful green tunnel of ancientness.

Looking back over the development of my ideas about garden design, I recognize that the older I get, the more confident and relaxed I want my gardens to be (with the maturity that comes from not trying too hard), and the more I want to work with nature. This makes me think of a garden I visited as a child, with my mother. It was owned by a bohemian couple, a Mr and Mrs Alodise, who had taught at Clayesmore School in Dorset in the 1950s. There they had created wonderful theatre sets for the plays the children put on, and I was shown photographs of amazing angel wings covered with feathers. The garden was on the edge of chaos, with roses tumbling out of trees and asters growing wild in the grass. We drank Lapsang Souchong tea in a circle of short grass cut out of the meadow. They both wore broken straw hats, and, as they sat in the long grass, they were so at one with their environment that they were completely covered in butterflies.

At Allt-y-bela, I have tried to echo this relaxed confidence. The garden is full of roses – not in the form of a formal rose garden, which would look ridiculous in this landscape, but in the meadows and tumbling out of trees and hedges; and not just the ramblers with sprays of small flowers, as might be expected, but wonderful full-blown climbers like 'Madame Alfred Carrière'. In the borders I have trained them over 'rose domes' or pinned them down to cascade over the edges of walls. But even in the borders that might be more traditional – such as those in the cottage garden, where I mix fruit and flowers – I invite much closer engagement with the plants. The cobbled paths, which form a crisscross pattern starkly visible in winter, take you among the flowers. Such close inspection encourages me to be more careful with my choice of plants. I find myself seeking increasingly exquisite colours and forms, sometimes from newly bred cultivars such as the dark Solomon's seal (*Polygonatum* × *hybridum* 'Betberg'), and sometimes from historical specimens, such as rare florist's tulips.

In one of my favourite parts of the garden, the kitchen garden, raised oak beds contain a mixture of fruit and vegetables, with flowers for cutting and as companion plants. I love the element of pattern-making in the layout of the beds, and the chance to alter its appearance every year with the hazel

FOLDOUT
For the first few years at Allt-y-bela we added thousands and thousands of snowdrops and *Narcissus lobularis*, the small native narcissi that seeds itself prolifically when the conditions are right.

By early May the garden is filling with sumptuous colour. The early-flowering species peony 'Molly the Witch' (*Paeonia mlokosewitschii*; below) has perfect creamy-yellow globe-shaped flowers over glaucous leaves, and forms a stunning contrast to the strong, dark tulip 'Queen of Night' and the orange 'Prinses Irene' (left).

Within the courtyard garden I planted my favourite rose, 'Cardinal de Richelieu'. Although it flowers only once, I enjoy its fleeting glory.

Pleached crab-apple trees frame the courtyard garden, giving it a sense of enclosure. Under the canopy of standard box balls I grow my rarest, most special plants.

structures we build to support the plants. Pears are trained on arches over the paths to celebrate the arrival into the kitchen garden, and step-over apple trees edge the beds. This part of the garden is enclosed – it has to be, since I think we instinctively need at least the sense of protecting our crops from animals and the elements – but it is a low enclosure of hazel wattle on oak posts, which sits well against the fields beyond.

There is more fruit in an orchard on the other side of the garden. The trees there are young, and I plan to add layers of under-planting and structure. The garden will always be in a continuous state of development, but it's coming. There is a new nuttery, too, where I will put *Pulmonaria* under the trees – just in delicate white – and where I can plant my white *Aconitum* 'Ivorine', the woodland peony *P. obovata* var. *willmottiae* and white martagon lilies. In these areas, beyond the immediate garden, the intention is not to create borders, but rather to add diversity and richness to the landscape, and to combine gardening with husbandry. By contrast, a richer, more sophisticated palette of plants is beginning to inhabit the space closer to the house.

At Allt-y-bela, I have found that the changing of the seasons has a particularly dramatic effect on the way the garden is viewed and used. In winter the landscape consists of short grass and bare branches; there are few evergreen trees around – just a little bit of holly in the fields – and, interestingly, the garden shrinks. The layers of topiary around the house are the focus. They hold the eye, and in the cold and wet one doesn't want to explore the wider garden. As we move towards spring, it is as though a veil is lifted. In January the first snowdrops flower, and the eye is drawn out as far as the common, then all of a sudden more and more snowdrops come up beyond the woodland garden. The eye roams further, the clutches of winter loosen, the narcissi and hazel catkins appear, and the witch hazels planted among them start to flower, glowing with shades of cinnamon, ginger and fiery red. Then come the meadows of spring bulbs at the back of the house, resplendent with crocus, *Iris reticulata* and early species tulips.

In autumn and winter the grass at the back of the house is cut short and joins seamlessly with the grazed fields beyond, but by spring, all of a sudden the garden begins to fill up with flowers.

Long shadows are cast by the light of a summer morning on grass and cobbles. The copper-beech topiary is in its richest purple guise, enhancing the greens around. The ancient band of cobbles next to the house provides well-drained growing conditions, which accounts for the unexpected success of the self-sown French lavender.

A little later, as the soil warms up, the fields around the house are covered in daisies and buttercups, playing catch-up, and the link with the landscape is established once more. My own meadows are full of buttercups, too, but are enriched with species plants and unexpected cultivars, such as *Trollius* × *cultorum* 'Lemon Queen', *Centaurea* 'Jordy', *Geranium phaeum* and *Aquilegia vulgaris* 'William Guiness'. These are all originally meadow plants, but I'm also experimenting with cultivars, putting sanguisorbas and astrantias, even roses, in the wild grass. By the end of June it feels as though my flowery mead extends all the way to the high ridges around us, because the fields are full of wild flowers, including wonderful orchids. So, in a way, the garden isn't just 2 hectares – it reaches as far as the eye can see.

The moment I saw the garden, I said the house was like an exotic pearl sitting on a cushion of green velvet, and now we're embroidering the cushion with native and species plants. The topiary is the Elizabethan stumpwork on the cushion, and my rarities are the occasional golden threads that give it another dimension. It is all very delicately crafted, all hand-stitched.

The vegetable garden at the back of the house is on a gentle south-facing slope. It is enclosed with hazel-wattle fencing and sturdy oak posts. These are materials that grow around us, so the plot feels honest and 'of' the place, in keeping with the house without trying self-consciously to be a 'medieval' garden.

The full exuberance of June is tempered by the cool green of the fields beyond. The pale grey-lilac double *Geranium* 'Gernic' (Summer Skies) is the perfect link between the pink and magenta roses.

PAGES 46–47
A froth of ox-eye daisies links the meadow directly to the house. The combination of topiary and roses brings the necessary weight and structure to the garden, which can be seen only in the context of the landscape. A crisscross of box hedging holds the herbaceous planting in an asymmetric space at the back of the house. The lattice is repeated on the walls in the hazel rods that will soon be used to train the shade-tolerant pale-leafed *Rosa* 'Astra Desmond'.

ABOVE AND RIGHT
The common was a thick tangle of bramble, elder and bracken when we arrived, but, with a regime of cutting and cutting again, it has returned to meadow. The orchids are reappearing, along with the meadow buttercup, and I have made additions such as ragged robin and *Iris* 'Perry's Blue' (above), as well as more unexpected cultivars that I have 'liberated' from my borders, including *Aquilegia vulgaris* 'William Guiness' (top).

PAGES 50–51
Looking across the cottage garden it is clear how the earthworks of the natural theatre have balanced the garden. I will continue to add layers to the garden, but it seems as though the house now sits comfortably in its surroundings.

Essential
Roses

I use the rose – one of my favourite of all flowers – in three ways: to clothe buildings and soften walls; to add impact and weight in mixed borders with its blowsy, perfumed splendour; and in the wilder parts of the garden, bringing unexpected sophistication to a meadow or tumbling through the tree canopy in great frothy cascades. Whether a design is traditional or in a more modern style, roses are essential to my gardens and I always find a way to include them.

I love the habit and scent of old-fashioned roses, which bring associations of well-known gardens in which they have been grown before, but I often mix them with newly bred repeat-flowering varieties to extend the season. It's very exciting to discover new ones, such as the vigorous and long-flowering *Rosa* Sir Paul Smith ('Beaupaul'), its bright pink petals backed with a paler hue. At Allt-y-bela it tumbles over a wall, bringing glamour to an otherwise country-style palette, but it would also look strikingly contemporary in a city plot or in a Mediterranean garden, cascading out of an olive tree. I like to visit rose-growers in June and wander through their fields, seeing as many varieties as

OPPOSITE
With casual grace, *R.* 'Bobbie James' (top and middle) and 'Seagull' are encouraged along the iron railings at Llowes Court in Powys.

THIS PAGE, CLOCKWISE FROM TOP LEFT
R. helenae hangs in veils, while *R.* 'Paul's Himalayan Musk' is a fountain of flower in the old apple tree at Guanock House. At Allt-y-bela, *R.* 'Königin von Dänemark' is a rich treat in the meadow. At an Oxfordshire manor the scrambling *R.* 'Cécile Brunner' dresses down the formality of the yew hedge, and the same rose tumbles between border and meadow at Allt-y-bela.

possible and finding out about their health and habit, as well as discovering their true colours and scent, neither of which can be conveyed in a catalogue.

Subtlety of colour is very important in the borders I design, and I'm always aiming to portray a particular mood, using blended colour and texture. I particularly like the way roses bring a change of texture and scale to a border, especially if they are trained over the walls at the back or tied on to domes and pyramids made from hazel. Structures of this sort, wrapped with roses, are an excellent way to introduce a vertical element, especially since they stand in winter, like topiary, when all the herbaceous plants have died back. Training roses in this way gives a greater volume of flower in a much smaller space than a sprawling shrub would occupy, since by slowing the sap in the stem it encourages flowering shoots all the way along the branches, not just at the tips. Wall-training has the same effect.

RIGHT
Trained in a lattice on the walls of Haddon Hall, *R.* 'Queen of Bourbons' adds a vertical element at the back of the border.

OPPOSITE
Roses on hazel supports punctuate the border. For the Chelsea Flower Show in 2012, domes of tightly trained roses gave the impact and appearance of rose topiary, but even as looser shrubs, they add weight and depth of colour to a border.

FROM OPPOSITE, BOTTOM LEFT *R.* 'New Dawn'; *R.* 'Reine des Violettes'; *R.* 'Ausland' (Sceptr'd Isle); *R.* 'Blush Noisette'; *R.* 'Austiger' (Queen of Sweden); *R.* 'Fimbriata'; *R.* 'Ausdrawn' (The Generous Gardener).

I love the softening effect roses have on architecture, and I often choose the climbing roses 'New Dawn' and 'Albertine', pruning them with great precision and keeping them tight to a building. On other occasions I might choose something like *R.* 'Madame Alfred Carrière' and actually encourage the plant to send out long, drooping fountains of shoots, so that there is a great volume and it feels informal, almost as though it's tumbling off the building. On a north wall at Allt-y-bela I've used *R.* 'Astra Desmond', which does well in shade, and it more than compensates for its lack of scent with its leaves, which are an incredible fresh green, and small, shapely flowers that last for almost a month. The blooms open from grey-pink to a stunning pale cream with a dark eye. Its neat and shapely form is beautiful, but where I needed a vigorous, large rose for impact on a long boundary wall at a garden in Oxfordshire, I used the magnificent *R.* 'Cooperi', whose one dramatic flush of flowers is now a much anticipated, if transient, event.

OPPOSITE, CLOCKWISE FROM TOP LEFT
R. 'New Dawn'; *R.* 'Madame Alfred Carrière'; *R.* 'Cooperi'; *R. banksiae* 'Lutea'.

THIS PAGE, CLOCKWISE FROM TOP LEFT
R. 'Astra Desmond'; *R.* 'Climbing Iceberg'; *R.* 'Buff Beauty'; *R.* 'Aussaucer' (Evelyn).

Kingsmead Mill

At the front of the mill in the 'tail race', the water is completely calm as it flows away from the mill wheel, mirror-flat and tranquil.

The approach to Kingsmead Mill in Wiltshire is a journey of anticipation. The main drive leads over a low, arching bridge and through a tunnel of horse chestnuts, with the feel of parkland beyond. The house, with its unusual pale blue window frames and its air of a French mill, is set behind imposing eighteenth-century railings. It is arranged on many different levels, and the stone tiles of the lovely roofscape catch the light with each change of angle and direction. It has the look of a naive painting. This rambling mill house, although not cottage-like, is not overly grand either, and the overwhelming impression is that of a dwelling perfectly at one with its watery setting.

The garden, in all its detail and intensity, is contained within a relatively narrow strip of land at the front and back of the house, and this is part of an extensive site of beautiful water meadows, a racing river and mill chases, and an enormous millpond. It was clear to me that the presence of the water and the strength of the setting must be respected absolutely, and that little work – apart from some editing of trees and repairs to riverbanks – was needed in the outer parts of the site. Anything more would have been superfluous and could have upset the balance of the place.

This is a garden of extreme calm and energy, both of which come mainly from the play of water. Still, reflective streams sit motionless, while elsewhere water flow fast over weirs, creating rivulets and frothy white channels that thrash at the tips of the weeping willows before pouring into the enormous millpond. There, the water sits very calmly before it takes another route over yet more weirs. Of course that energy changes with the seasons: in the summer, when there has been little rain, it is very tranquil and still, and in the autumn and spring, when there is lots of rain, the water races through with ferocious energy.

Beyond the dashing water are quieter areas of pools and marshy ground, and a walk through a narrow avenue of ancient black poplars that soar cathedral-like to the sky. There is a feeling of peace among the thick, dark trunks, and with no sound but the wind in the trees it is very atmospheric. From here there are views into the fields beyond, where cattle graze on the old ridge and furrow, surrounded again by beautiful water meadows. Beneath the soft,

PREVIOUS PAGES
The river divides here and is controlled by sluice gates, which send water either dashing down this weir, or directly to the mill race.

The entrance to the front courtyard, with its elegant wrought-iron gates, is much later than the original mill buildings, but generously encloses the space, giving an appropriate sense of formality and scale for the garden within.

Yew is kept clipped tight and low to frame the formal beds with graphic precision. In contrast, the planting within this structure is allowed to billow, softening the edges.

green woodland canopy is an under-storey of wild garlic and cow parsley, and silver pools with great floating pads of green duckweed and highlights of bright yellow *Iris pseudacorus*. This part of the garden has a myriad of small paths, not paved, not hard landscaping, but simply worn out of the ground, feeling very natural. It appears remote, but every so often you turn a corner and get a glimpse of the house across the water, through the trees.

The house effectively sits on an island in the arms of the river, held in by the mill chase. This separation allowed us to create a more formal garden on the narrow strip adjacent to the buildings. So close to the house, the gardens do not compete with or distract from the more natural beauty beyond. At the front of the house, where the railings form an enclosed courtyard, the structure of the low, clipped yew hedges and a formal pool was already in place, but when I first visited the beds were full of standard roses, all growing out of bare soil, which felt sterile and rather municipal. I wanted to create a much more atmospheric planting, more suitable for a watermill – the sort of thing you would have seen at the turn of the twentieth century when somebody who lived there had a passion for roses and aquilegias.

We softened the structure, adding to the existing wisterias, and planted lots of the more relaxed old-fashioned roses, supported on discreet metal domes that come out of a rich tapestry of perennials. There are lupins, salvias, big oriental poppies, geraniums and aquilegias. They all come to a pinnacle of flower in June, when it feels as if the garden is *en fête*, and the underlying structure almost disappears. At other times of year this area is allowed to step out of the limelight. In winter the clipped evergreen structure forms a contrast with the natural structure in and around the ponds and in the water meadows, with their enormous crack willows (*Salix fragilis*) and black poplars.

On the other side of the house we wanted to use plants suitable to the water's edge, so there is *Iris* 'Perry's Blue', and astrantias and camassias that like the damp. The planting at the edge of the river was also designed to feel like a water meadow, so we included cow parsley and self-seeded aquilegias, to represent the plants that grow naturally outside the formal garden. In time I hope these aquilegias will

We turned what had previously been a garden of stiff, formal roses in bare earth into mixed herbaceous borders, with a much softer, more abundant feel.

OPPOSITE
Rosa 'Reine des Violettes' is combined with *Lupinus* 'Masterpiece', *Papaver orientale* 'Patty's Plum' and *Allium hollandicum* (syn. *aflatunense*).

CLOCKWISE FROM TOP LEFT
Aquilegia vulgaris var. *stellata* 'Black Barlow'; *Rosa* 'Cardinal de Richelieu'; *Geranium pratense* 'Mrs Kendall Clark'; *Papaver orientale* 'Patty's Plum'.

Iris 'Perry's Blue' (below) grows by a low stone wall in a bank of native grasses that slopes down to the water's edge. On the other side of the wall is the mauve flower of sweet rocket (*Hesperis matronalis*).

spread their wings a little and get carried down the stream to populate other areas.

Running parallel to the old stone wall at the back of the house was a laburnum tunnel that we wanted to keep. Either side of a path of granite setts is a bed filled with bulbs and hellebores for spring, and punctuated by box balls and the roses 'Madame Alfred Carrière' and 'Souvenir de la Malmaison' for fresh scent and pale blooms later in the year. This path leads to the end of the garden, but it left us with an odd wedge-shaped piece of ground that was tricky to work with. In order to create structure that related to the house, we put in a series of hedges running parallel and perpendicular to the building, and called it the Box Garden. The hedges are of hornbeam (a tree that can cope with the damp), clipped squarely at the top but varying in height and joining at the corners to form 'boxes'. The height and structure give the area a sense of scale, and also disguise the garden's triangular shape. Between these hedges, a path zigzags past ferns and martagon lilies growing in the shade of a pear to an oak landing stage at the end. A rowing boat, kept in the boathouse further upstream, picks passengers up from here before being gently rowed up and down the river, picnic aboard.

The extra enclosure provided by the hedge allowed me to plant this area with winter-flowering shrubs and large herbaceous perennials, without fear that they would appear too ornamental next to the river. But closer to the house, where a large retaining wall keeps flood-water out of the house, we created an oak walkway on top of the wall, and an area for sitting and eating, surrounded by a mass of planting that we positively wanted to link with the river. The simple palette of plants – *Iris* 'Perry's Blue', astrantia, camassia, sanguisorba, *Angelica sylvestris* 'Vicar's Mead' and *Centranthus ruber* 'Albus' – is designed to provide a link with the native water's-edge plants beyond. To give structure and to form a light screen between the oak terrace and the house, we added three stump willows, as another direct reference to the planting along the river's edge. Whereas before the river was hidden behind its wall, now the water's edge can be seen much more closely, and from one of the best aspects of the house. It is the most natural place to sit outside to eat, being so close to the kitchen. It looks on to the far banks

The silvery water in the man-made 'tail race' snakes away from the mill house, through a parkland of mature trees and mown grass.

The subtle colour of this cast-iron bench and the edge of grass that is allowed to grow long ensure that the garden feels right for its country setting, and not too smart.

On the other side of the garden, away from the house, the atmosphere is much wilder. A magnificent double row of black poplars (*Populus nigra*) lines a path connecting a large pond and a glade in the woodland.

of the river, where lush grass grows thickly; the stream and pond edges are natural banks of iris, and beyond are big poplars and willows.

The water all around the house brings an incredible quality of light to Kingsmead. It bounces up from the river and reflects on the underside of the canopy of trees, creating a magical sense of space. And there is a lot of space here. This is a garden that invites you to stride out, to explore the far reaches, to find the connections to the landscape beyond its boundaries and then return to the house at its centre. The garden takes on an entirely different atmosphere and meaning in autumn, as the golden leaves of willow and poplar gradually give way to the bare branches of winter. If anything, that is when the light is most effective, and when the contrast between the wild areas and the small cultivated garden is most evident. The domesticity of the clipped topiary and managed hedges, of the jewel-like winter bulbs and daphne flowers, is more human and precious in the context of the impressive force of nature represented by the river in full winter flood.

Native black poplars (opposite) are rare, and are found only in watery ground. Here they border a series of boggy areas and small ponds, colonized by the native yellow flag iris and cow parsley or Queen Anne's Lace (*Anthriscus sylvestris*).

THESE PAGES AND OVERLEAF

In the beds closest to the river to the west of the house are *Iris* 'Perry's Blue', *Camassia leichtlinii* 'Alba', *Sanguisorba menziesii*, *Astrantia major* 'Claret' and *A.* 'Roma'.

Looking back to the house, you see the newly planted pollarded willows, with just one year's leafy growth on their thick trunks. They will be left for three years before they are pollarded back to the knuckle once more. They separate the seating area (on the raised oak deck) from the house.

A second oak deck extends over the walled mill chase, with oak steps down to a landing stage where a boat can take on passengers for a trip up river.

Oxfordshire Manor

PREVIOUS PAGES
The courtyard garden has the feeling of an outside entrance hall, enclosed by the pleached crab apple on one side and a billowing yew hedge on the other. Lavender and ox-eye daisies grow out of the chippings in a modern 'knot' of box hedges.

RIGHT
A wide doorway leads into the kitchen garden, which has simple box-edged grass paths and is surrounded by ancient stone walls.

The first time I saw this manor house, it was completely shrouded in a heavy, dark veil of holly and enormous laurels. It felt cold, even sinister. The garden had no atmosphere, and no link with the house or with the landscape beyond. It was divided by big *Leylandii* hedges into numerous compartments, leaving no sense of the footprint of the site but rather creating the feeling that one might be on the edge of the neighbour's garden at any point.

Before we could do any designing, we set about editing. We removed the conifer hedges and the large windbreaks of laurel, and realized all of a sudden that this wonderful Norman manor house sat in an ancient, undulating landscape that had not been discernible before. Venerable beech trees lined the drive, and within the grounds were magnificent oak trees. There were exciting glimpses of the medieval park that this once must have been. Removing the hollies on the banks around the house exposed a moat that remained on three sides. It was satisfying to see the clean lines of grass going down to the water once more, as the house regained its status as a moated manor. The water isn't there all year round – during some summers it dries up entirely – but there is a wonderful excitement to be felt when it returns and the moat starts to fill. The atmosphere changes completely: suddenly the house is across the water once more, sitting on an island in a wonderful landscape of green undulations.

The house was built in 1174 and has Tudor additions, and, most recently, alterations undertaken in the 1920s by the architect Detmar Blow. The layering that was added in the 1920s was elegant and refined, the craftsmanship superb. The double-height bay windows added at that time let light into the darkly panelled rooms, and allow much better views of the garden than must have been possible before. The young family – my clients – who live there now make the house feel vibrant and modern. They have cleverly combined contemporary furnishings, furniture and art with the ancient Norman stone archways and floors and the Tudor interiors. I wanted the garden to reflect this layering of history: to have the atmosphere of a medieval manor-house garden, the sophistication of the 1920s and the excitement of a modern twist.

OPPOSITE
The driveway has been kept very simple. Soft verges are planted with native beech and loosely clipped box topiary among tulip 'Queen of Night' and cow parsley for the spring.

PREVIOUS PAGES

Differently shaped topiary adds scale and drama to the courtyard, but feels more casual than if it were all in the same form.

BELOW

A walled cutting garden is laid out very simply as a series of small, square beds that are easy to manage. Here they are filled with tulips for the spring.

The materials from which the house is built are typical of an Oxfordshire manor. The walls are mainly of a honey-coloured stone, apart from the lime render and oak panel above the porch, and the roof is of beautiful stone tiles. In the garden we wanted to use a palette of plants to create the perfect framework for a manor house, so the first thing we did was to decide on our main plants: yews as hedging and topiary, fruit trees, wild flowers, lavender and box. There was already a huge yew hedge, but it had been cut perfectly straight and felt rather suburban. With a great leap of faith, my client allowed us to cut into it with chainsaws and remodel it drastically. The great billowing, undulating hedge that we have now feels ageless and confident – at one with the house.

The approach to the house is a critical part of any garden design. Here we wanted it to be really atmospheric, and steeped in history. The house is at the centre of a village, and I felt it was important to celebrate its presence within the settlement by changing the gates to an open rail design in oak. The framework gives glimpses of a wonderful wooded driveway of beech and hazel, and of the long grass verges, which are lush with wild flowers. We have added box balls, some neatly clipped as a contrast to the wildness, and others allowed to grow more loosely. Early in the year there are carpets of snowdrops; often, when I am starting a design and there is no garden to speak of, I plant thousands – both as bulbs and 'in the green' (transplanted just after flowering) – to build up the layers that suggest a site has been gardened for years.

As you come up the drive, you turn a corner and the first thing you see is the beautiful old church. Then directly in front of you is the most striking facade of the house, with its enormous doorway and Norman arch. Before we started work, cars would pull right up to the house, blocking this crucial view and ruining many views out from the principal rooms. We wanted to move the cars away and create an area of garden in front of the house, but at the same time we needed to create access through the space to the parking at the back of the property. We decided to make a courtyard garden set with the same limestone chippings as the drive.

We first populated the limestone surface with anchor plants, topiary trees of varying shapes, which

Pots of tulips (the pale pink 'La Belle Époque' and 'Bruine Wimpel', the deep purple 'Paul Scherer' and the orange-red 'Cairo') and small bulbs such as hyacinths will be used on tables to decorate the garden.

feel less formal than a repeated single shape and are another way of giving the impression that a garden has evolved over time. Then we added another layer, a transparent boundary to define the courtyard space, in the form of a row of pleached crab-apple trees. We used a variety called *Malus × robusta* 'Red Sentinel' – which has very bright red, almost glitzy crab apples with that modern twist that matches the striking freshness of the interiors – rather than the little russet-coloured one, which would have been far too medieval. You can see through the trunks of the crab apples at ground level, but they frame the house. It was important that this boundary was made of fruit, because when it flowers in the spring the blossom is a big part of the essence of the garden.

Next we added crisscross layers of box hedging, which we clipped into low, cloud shapes. I wanted it to feel like an old knot garden that over time had dissolved and fallen apart, so the lines are not solid and there are bits of hedge missing. Within this framework we added lavender, which grows very well in the free-draining surface. You brush past them on the way to the door and get that quintessentially English scent. To soften the space even further we planted ox-eye daisies (*Leucanthemum vulgare*) among the chippings, echoing the ones we planted in the wild-flower meadows (see page 88). This has the effect of pulling the atmosphere of the medieval park into this part of the garden.

From the courtyard garden you can choose to take a door leading to the kitchen garden, or there is a narrow opening in the huge yew cloud hedge. This leads into a wonderful area, bounded by the moat, that we call the Box Garden. The lovely undulations in its grass are the result of digging out the moat: where the soil was piled up it sagged and slumped. On these gently rolling daisy lawns, we placed intersecting box hedges on a much larger scale than those in the courtyard garden, but in the same diamond lattice pattern. The odd table, seat or picnic rug can be placed in this lovely green space. Two gardens provided inspiration for this area. First was Garsington, where I had been to see the opera, and its wonderful Irish yews that come out of square-edged beds. I loved the verticality and character of that garden, and wanted to add some of that here. Second was a garden at a girls' prep school,

OPPOSITE
A sculpture by Emily Young, *The Lovers*, brings life and drama into a very green environment.

ABOVE
Outside the kitchen door, there is space for a chair and table. It's very important to be able to sit and enjoy the garden.

In the courtyard the rambling, organic nature of the clipped hedges contrasts with the linear, regular planting of the pleached crab apples.

Hanford House in Dorset, where my godmother was a teacher. The box garden there was simply a series of clipped box hedges with green, grassed spaces between, where I had many picnics as a child.

Against this side of the big yew hedge in our Oxfordshire garden we created one long herbaceous border. It evolves, and over the years we have changed the planting many times. As I write it has fresh lime greens, soft lilac blues and washed-out blues and pinks; in the spring it is planted with bulbs in lots of very bright pinks. The walls of the house are clothed with an amazing old wisteria, which we brought back to its full flowering potential through pruning and retraining.

Elsewhere is a large terrace with lavender planted in the cracks of the paving stones, a mirror image of the lavender growing in the chippings at the front of the house. Under the windows we planted lots of varieties of daphne, to perfume the house in spring with their atmospheric scent. There had been a fish pond at the centre of the terrace, but it felt superfluous next to the moat, so we filled it with soil and planted it with a clipped platform of copper beech.

On the other side of the moat we have kept things very simple. Recently we created brocaded wild-flower meadows, cut with patterns and planted with ribbons of green hedging. Behind the hedges, in the future, we'll plant very bright borders. So in a journey around the garden, you will walk through woodlands or across lawns and then enter these ribbons of hedges and discover a vibrant dahlia border or a rose garden full of wonderful colour combinations. But from the intense garden around the house, you will still look out across the moat to a subtle tapestry of greens and wild-flower meadows, and there will still be that feeling of a medieval hunting park. The fields belonging to the property, adjacent to the park, have retained an undulating ground pattern formed by the ridges and furrows of early farming. We added fruit trees on the ridges, making orchards that connect even more strongly to the history of the house.

On the other side of the house is the beautiful walled kitchen garden. Ancient apple trees, now very carefully pruned and rejuvenated, line the central path. Simple grass paths thick with daisies are edged with box hedging that was already there. All we

The ancient wisteria has been pruned and retrained, adding scented colour to the honeyed stone of the house.

In the spring the borders are full of bright-pink tulips planted among the herbaceous perennials.

Within the box hedges are spaces in which to put tables and chairs for casual, impromptu dining.

ABOVE AND RIGHT
A fresh palette of colour for the herbaceous perennial border: the lime-green of *Thalictrum flavum* subsp. *glaucum* mixes beautifully with *Aconitum* 'Stainless Steel'.

PAGES 94–95
The entrance to the kitchen garden is clothed with climbing *Rosa* 'Bloomfield Abundance' and framed with loosely clipped box domes.

needed to do in this garden was to replant vegetables and put in new fruit trees against the walls, to keep it true to its heritage. Under the apples we added a very gentle, uncomplicated planting, not too 'designed', with all the things I remember from walking around Cranborne Manor in Dorset as a child: brunneras with the peony 'Molly the Witch' (*Paeonia mlokosewitschii*) and wallflowers, campanulas and euphorbias – a spring explosion of colour. Then, in the side borders, ornamental artichokes are mixed with vegetables and sage plants with dahlias. The big dahlia border is influenced by the dahlias at nearby Rousham, and forms a true kitchen-garden riot of colour.

We reused the beds in the kitchen garden, keeping the ancient fruit trees and adding a layer of roses tumbling over the walls. Annuals, including poppies, are allowed to self-seed, giving a country feel.

ABOVE
In the areas used for vegetables, young runner beans climb up traditional wigwams made from hazel.

LEFT
Even in the kitchen garden there are seating areas for enjoying the mixture of flowers and vegetables.

OVERLEAF
A transparent layer of crab apple 'Red Sentinel' provides winter colour all the way through to February or March, reinforcing the garden's atmosphere of productivity.

Essential
Kitchen Gardens

One of my favourite parts of any garden is the productive kitchen garden. I particularly love the requirement for honest simplicity and functional design, but I also enjoy the process of adding layers of detail and planting to create a more pleasing environment. Devising a scheme of companion planting, adding step-over apple edging or delineating an entrance or path with standard gooseberry trees does not change, but rather adds to the essence of a productive garden. The secret of making a successful kitchen garden is to create a practical working space.

Where there is space, a kitchen garden can combine ornamental and productive planting. Box and yew topiary gives vertical scale to this beautiful walled garden. The brick-edged beds set in paths of limestone chippings are both practical and elegant.

Crop rotation is an important aspect of the kitchen garden, and has a great impact on the design. Whether it is scaled up or down, the division of the space into four is the most satisfactory starting point. Old walled gardens are invariably set out in this way. Finding one is rather like discovering the secret garden of Frances Hodgson Burnett's story; it doesn't happen very often these days, but I can remember as a child discovering these magical worlds that had fallen asleep. The structure was often still there, to be discovered under brambles and nettles.

Large walled Victorian kitchen gardens are particularly special, but of course not every garden can accommodate a plot of that sort. The sense of enclosure, though, is what makes them special, and that can be created whether you have just a few square metres or a couple of hectares. I think our response to the peace and tranquillity of enclosed spaces is deeply ingrained. The comfort of a walled kitchen garden stems from the primal knowledge that food plants need protection: early humans had to protect their food in order to survive.

One of the first things I do with a vegetable garden, therefore, is to give it a sense of enclosure. This can involve anything from very tall brick, stone or cob walls to a simple espaliered fruit hedge or a hornbeam hedge, or even estate railings with rabbit-proof netting. What is vital to me is the ceremony of stepping into that enclosed garden through a gate or door that can be closed behind you.

After enclosing the whole space we turn to the pattern and the holding in of the beds. Within the main division of the space into quarters we can create subdivisions by enclosing beds with low hedges or step-over apple trees, building up layers of texture and height and delineating pattern. It is not always necessary to maximize growing space, because self-sufficiency is generally not the object. Instead, I usually aim for vegetable gardens that are like delicatessen counters, with just the best things, grown for fabulous flavour. I might grow just a few earlies for the intense enjoyment of fresh new potatoes, for example, rather than trying to supply potatoes for the whole year. Similarly, with onions, I like to have some fresh and a few to store, but that's all. For me, it's about growing modest amounts; the choice of *what* to grow is most important. I plan crops that span the seasons,

If possible it is good to start with the vegetable garden. Whether it's in an established kitchen garden or a new area, the speed of the results is a great encouragement. The freshness and intense flavour of the produce, even that from the tiniest of gardens growing only herbs and salads, is always an incentive to grow more.

so there is always something tasty to eat, but not necessarily in great amounts.

Fruit and cut flowers are also part of the kitchen garden. It is important to have flowers like calendula and nasturtiums to bring all the benefits of companion planting, to protect against pests and diseases, as well as adding colour.

I particularly enjoy the fact that the vegetable garden is different every year. There will be gluts and disasters, new discoveries and disappointments, but that constant learning is what makes it interesting. You can try a new variety and find that it has the most perfect flavour, or suddenly discover that the tips of broad beans are delicious beyond words, so you need to grow more of them. This is a very personal, passionate part of the garden.

Wherever possible in my design projects I place a greenhouse and cold frames in the kitchen garden. This allows you to start your own plants, and to extend the growing season, for example by growing winter salads or ripening tomatoes a month or two before those kept outside. Terracotta forcing pots produce decadent, sweet pink rhubarb or blanched seakale, and traditional glass cloches allow you to bring on early lettuce in the ground and protect all sorts of young seedlings. These objects are all part of the husbandry, the deeply satisfying craft of vegetable gardening, and their beauty brings something extra to the aesthetics of gardening, as well as to its ceremony.

Horticulture, husbandry and craftsmanship come together in the kitchen garden, and all are vital for an aesthetically pleasing and successful garden. Seedlings are raised and nurtured until it is time for cropping. Hazel is grown and coppiced for beanpoles and peasticks, at which point dexterity and an artistic eye are required for the making of structures to support plants and protect crops. The gardeners shown here are (clockwise from top left) Will Smithson of South Wood Farm, Benjamin Pope of Queen Anne House and Scott McDermott of East Hampton.

East Hampton

The cobbled path between the guest house and the main house is sown with thyme, forming a carpet that wears through to the cobbles where people walk. In summer it is pinky purple with flower, but in the green it softens the distinction between the lawn and the hard surface.

I was thrilled at the thought of working on my first project in America, this beach house in East Hampton, on Long Island in the state of New York, and particularly relished the fresh challenge of a different environment and climate. In the summer it is wonderfully, reliably hot, and in the winter it is properly cold and there is a great deal of snow.

In fact, on my first visit the site was completely covered in snow. It was a sunny day, the ocean was crisp and blue, and the snow went right over the beach. It was useful to see it that way – a little like looking at a black-and-white photo; there were fewer distractions and I could get a sense of the scale. Very little was left in the garden: an apple tree or two, some ornamental cherries, a few big yew trees and box balls, and a couple of ancient *Magnolia* × *soulangeana*. These we kept, but it was just outside the garden boundaries that I found the inspiration for my design and for the planting.

At the side of the road in this part of the country asters and amazing grasses grow wild. Butterfly weed (*Asclepias tuberosa*), sea lavender (*Limonium*), artemisia and goldenrod grow out of the sand dunes, and in the wooded areas there are masses of native orange tiger lilies (*Lilium columbianum*). From the outset I knew that I wanted to use native plants, so that the garden would feel part of its environment. Interestingly enough, though, these local plants were not unfamiliar to me: many of the garden plants we use in Britain actually originate in America.

As well as using native plants, which I knew would perform well and feel right, I hoped to design a garden that held hands with its locality, by echoing the local agricultural landscape. I wanted it to be quite different from other gardens in the area, a little more contemporary, but still keeping traditional elements. I had never been to Long Island before, and driving out of Manhattan, leaving the hubbub and franticness of the city behind, I was struck by the way Long Island unfolds along the long, straight road. It's like a conveyor belt that delivers you to the Hamptons, and along the way you start to notice different things. At first there are pine and oak woods, and the ocean is invisible; at places like Bridge Hampton and South Hampton the landscape is surprisingly agricultural. There are pumpkin and strawberry fields, orchards growing apples, nectarines

PREVIOUS PAGES
The view of native seaside goldenrod (*Solidago sempervirens*) among the dune grasses at the edge of the beach in front of the house inspired my use of flowers and grasses in the planting around the house.

I love the agricultural language of the vineyards in the Hamptons, and used them as inspiration to create structure in the garden. The stretches of sand on the beach side of the garden provided another reference. I constantly remind myself of the location of the site on which I am working, since I believe it's essential to create a garden that doesn't feel in any way alien to its setting.

OVERLEAF
Either side of the cobbled walk, a sense of the local agriculture is reflected in the orchard trees, the vines and the kitchen garden beds, from which pumpkins sprawl across the paths.

An orchard of crab apples sit on the high points of the rolling, ridge-and-furrow lawn, and is punctuated by loosely clipped box. The crab apples are *Malus* × *zumi* var. *calocarpa* and *M.* Sugar Tyme ('Sutyzam').

and peaches, vineyards, potato fields and many organic vegetable growers. Beach houses sit among these potato fields, right next to little stalls selling impressive pumpkins. I really wanted my garden here to have a strong connection to this surrounding agriculture.

I also noticed that in the villages incredible copper and green beech trees have been sheared into huge topiary up to 10 metres high, much larger than I'd ever seen before. They are hundreds of years old and have colossal gnarled trunks that feel almost archaeological. They form clipped green architecture among the softness of the mature woodlands of unclipped beech and elm, acting like connectors to history. I knew they were another essential ingredient for the garden.

The positioning of the new house and guest house (nothing of the original buildings remained on the site) gave me the divisions of the garden and the chance to develop it as a progression through three distinct areas. The first is a very lush, green woodland environment around a courtyard at the entrance. This is the furthest point from the ocean, and is sheltered from scorching winds and salt spray, so big trees can grow here. They provide the shade that is so important in the heat of summer, and the opportunity for under-planting with lots of native spring perennials like the Virginia bluebell (*Mertensia virginica*) and carpets of *Vinca minor*, below deciduous shrubs such as *Cornus*, and the fringe tree (*Chionanthus virginicus*).

From the woodland entrance, a drive of crushed oyster shells meanders past the guest house through fruit trees. The snaking of this drive forces the visitor to slow down, and the feeling of driving through an orchard gives a connection to the agricultural language of this central part of the garden. We wanted to keep hard landscaping to a minimum and to use local materials. We used crushed oyster shells or marl for many of the pathways as well as the drive, because it feels closely connected to the beach. Some of the paths nearer the house are of red brick, like its chimneys, and the main path straight through the centre of the garden is made of river-washed cobbles from a local quarry. We saw the same cobbles in the ground when the house's foundations were being dug. The laying of each pebble, done carefully by hand, created an

The 'ridge-and-furrow' lawn was an influence that I brought with me from the ancient English landscape, but I feel it also echoes the movement of the ocean.

Reclaimed fence posts with rusted wire were used for training antique vines. Such materials help to give the new garden an instant sense of age and a link to its location.

Among the vegetables, the lime-green *Zinnia* 'Envy' and *Tagetes* work well with the leafy crops, and can be cut for the house.

Strips of lawn are divided by paths of crushed shell and lines of wires supporting the vines, which are under-planted with strawberries. The box-edged vegetable beds cut through the lines in a sequence of rectangles.

extraordinarily beautiful texture, and the labour involved brought a sense of permanence to this new structure. To add to that we sowed creeping thyme in the marl and sand that fill the voids between the stones. In the summer, the thyme flowers and forms a contemporary purple-pink ribbon that links the main house with the guest house.

This path travels through the largest section of the garden. On one side, near the guest house, we created an orchard set out in a grid. It is deliberately not a complete grid: I plucked some trees out so that it looks like an old orchard, where some of the trees have died over time. I chose to plant crab apples, which hang on to their fruit all winter, attracting birds and other wildlife. The trees are planted on subtle ridges, inspired by the agricultural practice of ridge and furrow. The ridges are only about 60 centimetres high, but that is enough for them to cast shadows and create movement when the sun comes through. This stops the large area of grass from becoming flat and sterile, like a golf course. Spring bulbs such as *Fritillaria meleagris* are planted in the dips, and *Chionodoxa* and lots of *Crocus tommasinianus* and *C.* 'Prins Claus' on the ridges. Later in the year it is just grass, cut slightly longer than the lawn.

At the centre of the orchard a double row of vines subdivides the grid and, from the drive, gives a wonderful view that will eventually frame a shade pavilion or garden house. We planted 'antique' vines, so that they immediately connect in the mind with those in the vineyards on the journey from New York. We trained them on steel wires that are allowed to go rusty and even to sag, supported by old wooden fence posts that I found piled up at the landscaper's yard. These posts (which have holes where the horizontal wooden bars would have gone) and the crisscrossing of threaded wire add to the pattern-making and agricultural language of this part of the garden.

Among the vines are smaller marl paths, and three beds of vegetables are edged in low clipped box. Here we grow a mixture of vegetables, and flowers for the house; not many, just complementary plants that bring colour to the vegetable beds: lime-green zinnias, orange marigolds and lots of tomatoes. The local growers have an astonishing range of heritage tomatoes, which are now grown

We planted ancient 'sheared' beeches, which brought great maturity to the garden. The trees are even more magical viewed from within the canopy, where you can stand among the sculptural branches.

The newly planted beeches, like giant topiary with their great domes, give a great sense of scale to the garden.

We sowed thyme seeds into the surface of crushed shells that beds in the top level of the round pebbles.

in the garden alongside herbs for the kitchen, and aubergines, peppers, chillies and pumpkins. The pumpkins spill out of the beds and inhabit the marl paths, mimicking the sprawl of pumpkin vines in the fields near by.

A very big, plain lawn gives a breathing space between here and the main house. In this calm environment between the two buildings on the site are three enormous sheared beech trees. We were very lucky that the landscaper had these ancient trees for sale in its nursery. Their scale immediately links the two houses, and allows the garden to feel rooted in history. They are real characters, like wonderful old friends in the garden. You can walk right inside the canopy, and find yourself in a network of gnarled branches, looking out at the tapestry of green leaves, and seeing through to the next sheared beech and the next. It is like walking into ancient yew topiary or hedges in England. They emphasize the green palette of this part of the garden, and – especially important in the winter – they give the garden vertical structure.

In the winter all this structure is very visible, but in the summer the garden becomes very soft. Closer to the main house, the paths run through a veil of meadow-like planting, like the froth of the ocean. When you're standing on the beach side of the house you see waves crashing in the distance, and sometimes the air is full of sea spray, which creates an amazing quality of light. I wanted the planted beds to reflect that, so the planting is free and loose, in the style of a native American prairie. There are grasses such as *Briza media* and *Bouteloua gracilis* 'Blonde Ambition', with its caterpillar-like heads that act as brilliant light-catchers, lots of native asters, *Echinacea tenesseensis* 'Rocky Top', *Agastache* 'Pink Haze', and *Verbena stricta* and *V. bonariensis*, all with small heads that seem to float, and again catch the light.

Finally, next to the house, a slightly raised courtyard is held in by espaliered apple trees bearing bright red fruit (*Malus domestica* 'Gravenstein' and 'Gala'). Out of the crushed-shell paths grow plants that are allowed to seed around: *Origanum laevigatum* 'Herrenhausen', *Calamintha nepeta* and more *Verbena bonariensis*, and for the autumn, against the shingle walls of the house, white Japanese anemones for freshness. This loosely planted mix surrounds the

The creeping thyme 'carpet' is soft and rich in flower. The varieties are *Thymus praecox*, *T. coccineus*, *T. quinquecostatus ibukiensis* and *T.* 'Magic Carpet'.

Planted in the crushed oyster-shell surface of the paths, a variety of perennials give the softness that is normally associated with self-seeded plants.

Box and magnolia trees create a green plinth for the newly built house.

Contemporary lead containers are planted with *Salvia* 'Amistad' and *Pelargonium tomentosum*, whose leaves give off a strong peppermint scent as you brush past them to enter the house.

An antique iron bench is placed in the shade of a *Magnolia × soulangeana*, in the courtyard garden.

The brick pathway leads through espaliered apples into a meadow-like border of native grasses and perennials, making a connection with the local, natural flora.

three most formal beds in the garden, which are framed with exaggeratedly wide, low box hedges. These big beds give us the opportunity to place more formal plants, such as roses and peonies, and act as a plinth for the ancient magnolia, but the planting is still softened by the native grasses and plants, such as *Amsonia illustris*. The whole area is under-planted with wild strawberry, which forms a matrix for the other plants to grow through, both here and elsewhere in the garden – anywhere we needed to cover bare soil.

The ocean side of the garden is completely different. The house looks down on two or three large ridges of sand dunes, with a wonderful view of the beach and ocean beyond. This part of the garden had to be uncomplicated. There is a very visible change in the number of species that grow naturally close to the ocean: on the beach there are only beach grasses, goldenrods, a few sea lavenders, *Artemisias* and beach peas (*Lathyrus japonicus*); inland it becomes a little more protected from the salt winds, with bayberry (*Myrica*) and then beach plums (*Prunus maritima*) and native cedars. Further inland still there are scrub oaks (*Quercus virginiana*). It was apparent that this was the required planting language on this side of the garden. There are no box hedges, and where previously there had been cultivated grass running right down to the dunes, we chose to extend the sand and beach grasses upwards, much closer to the house, using lawn only on the flat plateau immediately next to the house.

Close to the swimming pool and main house are two big square beds, where I created ribbons of clipped bayberry. This is a modern knot garden, a parterre of beach plants, with native grasses such as switch grass (*Panicum virgatum*) and little bluestem (*Schizachyrium scoparium*), sea lavenders, sea holly (*Eryngium planum*) and more butterfly weed, with its orange flower. For extra height, big bayberries with sculpted stems rise out of the planting, and the beds are mulched with crushed seashells. It all connects with what grows wild on the dunes: I wanted nothing to jar as the eye travelled out of the garden to the magnificent horizon beyond.

Nectar plants such as *Verbena bonariensis* (top) and *Agastache* 'Pink Haze' have been very effective at attracting butterflies to the garden.

The borders of grasses and native perennials feel very natural and bring a sense of the sand dunes' mix of flowers and grass to this side of the house. The aster seen below is *Aster linariifolius*.

OVERLEAF
Leading from the garden down to the dunes, where previously there had been a alien-looking mown lawn, I felt it was important to re-create the sand dune, bringing it right up to the garden and planting it with native beach grasses and bayberries.

As you walk through the sand dune towards the house, the garden becomes more contemporary and formal, with plaits of bayberries in-filled with native switch grass, salvias and sea lavenders.

PAGES 130–31
The pool, with its simple lines, reflects beautifully the windswept bayberries that add a vertical, sculptural element to an otherwise horizontal landscape.

The Mill House

LEFT AND OPPOSITE
The front of the house is covered in roses – 'Albertine' and 'Zéphirine Drouhin' – with lavender and rosemary in the beds below.

The site of Lower Heyford Mill in Oxfordshire is very wet: a canal runs past, and a mill chase meets the river on the other side of the site. This leaves a linear garden between the two bodies of water. This marvellous setting, with water running past the house, brings great energy and a constant feeling of movement, but when I first saw the garden there were too many visual distractions and it did not feel at one with the landscape. A great deal of my design philosophy and approach stems from what I love, and from what I have absorbed over the years at my favourite garden, Rousham (not far from Lower Heyford). With images in my mind of how well the planting fits the setting at Rousham, when I first visited the mill it was obvious that many elements of the garden were very alien. It was interesting to discover that the awkward elements were those that fought with the energy of the water to make the garden static, as though holding it and weighing it down.

There was a lot of inappropriate planting, and the borders were at an odd scale. Elements such as the swimming pool appeared to have been dropped in from space, and were squeezed in on awkward levels. Where a flat area of land was needed for a tennis court, the soil had been simply scraped away; it stayed where it landed and was planted with species that were inappropriate for the site. Unnatural-looking Portuguese laurels clipped into large, heavy balls lined a track next to the river, blocking the view to the water.

At the back of the house, the lawn was about 1.5 metres above the internal floor, and reached almost to the wall of the house, with just a little retaining wall and some steps. Where anyone would want to spill out through the back door from the kitchen, therefore, it felt hemmed in and claustrophobic, like being in a dark, damp hole. To solve this, it was very important to me that the retaining wall be pushed back and the lawn lowered. There were fixed points – such as an established hedge and beautiful lime and walnut trees – that we wanted to retain, so to keep them at their current level and hold up the swimming pool we created a series of sinuously sculpted grass banks. This allowed us to drop the lawn by as much as 1 metre in places, letting a great deal more light into the house

PAGES 132–33
The pale pink rose 'Albertine' and darker 'Zéphirine Drouhin' are kept neatly trained to the walls.

through the back windows. We built a large terrace in an inviting spot for breakfasting in the morning sun, and, because it is just outside the kitchen, it is also used for other meals. A team of drystone wallers built magnificent retaining walls of Oxfordshire stone to keep up the new levels, and added four wide steps to the lawn. Suddenly house, river and garden were connected, and the space began to flow.

On the new terrace we wove delicate plaits of low box hedging into an intricate knot that wraps around the seating area. Within the hedges are violas and hellebores, among other plants, and some roses. On either side of the steps a pair of yew trees, clipped into large balls, dominate and contain the area, and separates the terrace from the green of the lawn. This relaxed, imperfect lawn, dusted with daisies and clover, runs almost to the edge of the river; there the grass is allowed to grow longer and a simple line of estate railings marks the river boundary. Beyond the terrace, therefore, the garden at the back of the house is minimal, in contrast with the abundance of roses on the house wall itself. The simple lawn echoes the meadows beyond the garden, which are grazed by horses.

We could not move the tennis court or the swimming pool, but we could establish a better visual link with the river. Impressive but towering willow trees, which had not been pollarded in recent times, cast deep shade and closed out the view of the river. Pollarding them put everything back into scale, restoring the house as the dominant feature and allowing it to sit much more happily within the garden. We took out all the inappropriate laurels on the raised grass walk next to the river, and in the woodland, restoring the house's breathing space. We left the native ground cover undisturbed, except at the edge of the wooded areas, where we added brunnera and aquilegia to grow wild in the grass. Now the garden once more allows views of the surrounding water meadows, which flood regularly, changing the landscape suddenly and dramatically. Huge lakes surround the house for a time, bringing in extra reflected light; when the water recedes, lush green is left behind.

There is a good courtyard at the front of the house, but the parking came right up to the front door and a wishing well sat in the centre. The house's elegant Georgian fenestration feels very

OPPOSITE
The pale blooms of *Rosa* 'Albertine' create a stunning, scented tapestry on the wall of the house.

BELOW
In the box-edged parterre in front of the house the seed heads of *Allium hollandicum* (syn. *aflatunense*) are left until the lavender flowers.

Planted into the chippings is a mix of *Dianthus carthusianorum*, with its small, bright pink flowers held on tall stems, *Pulsatilla vulgaris*, *Verbascum chaixii* 'Album' and *Lavandula angustifolia*.

At the side of the front courtyard, wisteria and a lead trough of lavender soften the view of the outbuildings, which are now used for accommodation.

simple and clean, and I wanted to match the facade with an uncluttered formal garden in front, to push the cars away. The building was once a working mill, and with the design of the garden we had to establish a hierarchy among the many doors that punctuate its front. The planting would have to turn a working building into a home. I designed a rectangular insertion in the courtyard, with triangular, box-edged beds at each corner leaving a central lozenge-shaped lawn. Pyramids of yew that line up with the main door mark the entrances to the lawn. To bring out the Provençal feel of the house – which in fact had a strong influence on the design of the garden as a whole – we planted nothing but lavender and alliums inside the box hedging.

Beyond the green, we united the whole of the rest of the courtyard with limestone chippings from a local quarry, which means that there are no hard lines between the drive, the parking area (now to the south, by the stable building) and the courtyard. We planted verbascums, dianthus and pulsatillas, and also lavenders that are so happy in the conditions provided by the chippings that they seed themselves around, reinforcing the self-seeded effect we were aiming for when we first placed the plants. The softness of the free-draining, free-flowing garden that inhabits the limestone chippings dresses down the formality of the crisp, green, enclosed lawn and formal topiary.

To hide the parked cars from the house, and create a different scale and texture from the low formal garden, we added three pollarded lime trees. They felt particularly appropriate for this garden, where the language of pollarding is very strong. (It is necessary to pollard the willows on the riverbanks, because as they grow they become unstable, and if they blow over they pull the riverbank up with them; traditionally they were also pollarded so that wood could be harvested for baskets, fencing hurdles and so on.)

Beyond the courtyard we made a small, enclosed kitchen garden out of a concrete-paved storage space where a horsebox was formerly parked. The new garden is shaded by a very large weeping ash that we did not want to lose; it screens the neighbouring houses and gives great privacy. Despite its shadiness, however, this is a very successful, intricate, intense vegetable garden. It is entered through a heavy oak

At the back of the house, the banks of the mill chase are full of meadowsweet (*Filipendula ulmaria*) and red campion (*Silene dioica*).

In the front courtyard we planted a cork oak (*Quercus suber*), which added to the slightly French feeling of that side of the house.

gate, and candelabra-espaliered pears line the path to a potager of chipping paths and raised oak-edged beds. There are salads, herbs and a few new potatoes, and to one side we built a free-standing white-painted wooden greenhouse.

Off the courtyard, to the side of the kitchen garden, a little serpentine path of limestone chippings leads to the north through a plantation of purple hazel (*Corylus maxima* 'Purpurea'). The stunning deep-purple leaves contrast with the greens all around, and provide a rich foil for a shady woodland under-planting of *Anemone sylvestris*, cyclamen, ferns, Solomon's seal, hellebore and *Sarcococca*. The path leads to a simple oak bridge over the canalized mill chase, and over it to a partially enclosed walled garden.

This magical, sunny spot is bordered on the south side by the mill chase, which runs under the main building through the mill wheel, and emerges from an old wooden structure at the eastern end of this walled enclosure. We made this the main herbaceous garden, but a margin of mown grass at the water's edge allows feet to be dangled in the stream on a hot summer's day. Beyond the grass

In early June (above) *Pulsatilla vulgaris* is still holding on to its wonderful seed heads, while *Scabiosa columbaria* 'Nana' (left) is in full flower, enjoying the well-drained conditions of the limestone chippings.

are deep double herbaceous borders, backed by a beautiful stone wall. From here the river and large millpond are visible beyond the walls, and terrific energy comes from the water rushing all around, but the walls give enough enclosure to allow exuberant and luxuriant planting.

The ancient weeping ash, which feels like the finale to the garden, stands at the very end of this space, and we also kept the golden Irish yews that brought in a glow almost like sunlight. For the many roses and perennials, we had to choose colours that work with the yews, resulting in a cool, somewhat Edwardian palette of white, icy blue and silver. A froth of *Geranium pratense* 'Mrs Kendall Clark' and *Nepeta racemosa* 'Walker's Low' is punctuated with foxgloves, bearded and flag irises, *Crambe cordifolia*, *Campanula lactiflora* 'Loddon Anna' and the creamy-yellow *Hemerocallis* 'Gentle Shepherd'. The presence of water means that the garden never feels hot and there is always a breeze, so it was very important to reflect that with the choice of hues. Hot colours would have jarred with the tranquillity, but the serene whites and pale blues are icy and cool against the backdrop of stump willows. We added more willows, to reinforce their place in the garden. They are allowed to grow for three or four years before being pollarded, and their knotty, gnarly stumps form the dominant planting in the garden, their silvery-grey leaves reflecting light and echoing the movement of the water.

The cool colour combination of *Viola cornuta* and *V. cornuta* Alba Group scrambling around the box plaits at the back of the house is matched by the sweet rocket (*Hesperis matronalis* var. *albiflora*) and *Nepeta racemosa* 'Walker's Low' in the herbaceous border.

The clump-forming plume thistle *Cirsium rivulare* 'Atropurpureum' gives an eye-catching spot of dark red among the cool blooms of *Geranium pratense* 'Mrs Kendall Clark', which is threaded through the herbaceous border.

I like roses best when they are planted in herbaceous borders, in combination with other plants. Here, the dark red of *Cirsium rivulare* 'Atropurpureum' is a lovely complement to the soft pink of *Rosa* 'Anna Pavlova'.

OVERLEAF
The light of late afternoon catches the white rose 'Auscat' (Winchester Cathedral) and a cloud of tiny white flowers of *Crambe cordifolia* in the full June flush of the herbaceous border.

Essential
Borders

Borders – exciting, ever-changing – are the most expressive part of my gardens. They can bring a feeling of drama or of tranquillity; they can be soft, romantic and old-fashioned or strikingly modern, with a precise balance of colour. The process of building a garden starts with the structure of hedges and walls, paths and terraces, but the atmosphere comes when the plants start to inhabit the space.

The two photographs on the page opposite show a double border with a very formal edge of wide stone. The plants are allowed to flop on to it in a relaxed way, but the strength of the structure

OPPOSITE, FAR LEFT
Astrantia major 'Claret' with the white flowers of *Anthriscus sylvestris* 'Ravenswing'.

OPPOSITE AND LEFT
The pale pink *Rosa* The Generous Gardener ('Ausdrawn') is repeated throughout these borders, with purple *Nepeta racemosa* 'Walker's Low' and the silver of *Artemisia absinthium* 'Lambrook Silver'. At the back, held high on tall stems, are the white flowers of *Valeriana officinalis*, with the mauve spikes of *Delphinium* Dusky Maidens Group, and at the front of the more formal border is the claret *Penstemon* 'Blackbird'.

and width of the path maintain the feeling of formality. The red pavilion was the starting point for the palette of plants, and I added lots of silver and purples and pale pink roses to balance the heat of that red. In the same garden (pictured on page 148), similar planting is used in a much more informal border, where the path is barely visible in summer. As you walk through you can touch and smell the plants, resulting in a completely different experience.

I start designing by imagining the mood I want to create, taking into account the way the border relates to the exterior and interior of the house. If it is close to the house, I might extend the season with tulips for spring and a few winter-flowering shrubs, but I want the excitement of a full, blowsy border in summer; as long as they are kept tidy, I'm happy to allow borders to brood quietly over winter, storing up their energy for a spectacular summer show.

The old walled garden shown on these pages has thin, chalky soil and bakes in the heat of summer, so pale, washed-out colours feel appropriate. The grass paths are edged with lavender; pale *Monarda* 'Elsie's Lavender' is massed behind it; and peppered among the pale swathes are the dark, nodding heads of *Aquilegia vulgaris* var. *stellata* 'Black Barlow', a contrast that makes the planting as a whole look even softer.

My borders are planned meticulously – with foliage and flower shape and colour as well as timing all considered – but I like the fact that, in the end, they cannot be controlled. They inevitably change over time: one plant will die out, and another will become too dominant; the aquilegias will seed and the alliums multiply faster than they can be pulled out. As gardeners, we manage the balance of plants by lifting and dividing them, and replenishing others, but the transience of the whole thing becomes part of the enjoyment. If, after a few years, we decide to move most of the plants to another part of the garden, what was a cool, sophisticated pale blue-and-white border can turn into an energizing display of fiery colour.

Tulips are a highlight of spring, and summer is of course fresh and abundant, but I particularly love autumn borders, because of the low light that picks up the colours so beautifully. At that point in the growing season, the asters that have stood politely

The pale lilacs and pinks in this walled garden include *Lavandula × intermedia* 'Grappenhall' (opposite, top left and middle); *Veronicastrum virginicum* 'Spring Dew' (opposite, top left); the perennial honesty, *Lunaria rediviva* (opposite, bottom); *Monarda* 'Elsie's Lavender' (opposite, top right; seen again opposite middle and with *Verbena hastata* f. *rosea* below left); *Aquilegia vulgaris* var. *stellata* 'Black Barlow' and *Geranium pratense* 'Mrs Kendall Clark' (below right).

by throughout the summer come into their full glory, and the borders are absolutely stuffed with flowers. For late summer at Cottesbrooke Hall in Northamptonshire I designed annual borders that have an extraordinarily ethereal quality against the permanence of the old brick walls (opposite). Every year the annuals and dahlias are planted out in early June and race, as only annuals can, to fill the space with their airy beauty. The narrow mix of colours gives the planting cohesion, but the contrasting flower forms and leaf texture jostle for attention and create a rich, opulent effect.

OPPOSITE
Spring (top left), summer (bottom) and autumn (top right) borders are very different in character; the colour palette can remain similar or change completely with the seasons.

THIS PAGE
The annual border at Cottesbrooke Hall. The mix of dahlias includes (from far left) 'Stolz von Berlin', 'Alloway Candy' and 'Elga'.

Bowling Green

PREVIOUS PAGES
Bowling Green sits high above Haddon Hall, with magnificent open views across the Derbyshire Dales: green deciduous trees; vibrant mixed hedges of hawthorn, buckthorn, elderflower and alder; old ridge-and-furrow fields; and dignified stone walls.

ABOVE
We did not want to disturb the gates that lead out into the park. They are a little crooked, and the pillars are cracked, but we liked the romance of decay. Perhaps one day they will be carefully repaired, but for now they form a fine contrast with the vigour of the grass borders below.

Travelling to Bowling Green, part of the Haddon Hall estate in Derbyshire, means crossing the broad, rolling hills of the Derbyshire Dales and dropping down into the little market town of Bakewell. Just beyond there lies the Hall. Recently I've been working on the Hall itself (see page 170) and on the Peacock Hotel, also part of the estate, but this was the first garden I designed at Haddon. Bowling Green was the Hall's games pavilion, where the family would go for entertainment and to play badminton and, of course, bowls; the building was extended in Georgian times, and became a house. This is the private side of Haddon, the place to which the family retreats when the Hall is open to the public.

The marvellous approach leads through the estate and climbs high above the Hall, delivering the visitor to one side of Bowling Green. Originally, those making the journey from the Hall would have come through the Eagle Tower and across the ancient parkland, through oaks and lime trees, to enter via a grand flight of steps and a pair of gates. Bowling Green is not as old as the Hall, but has a very strong connection to it, and I wanted to reflect and emphasize that link in my design, while also acknowledging its separate identity. It has a peculiar sense of playfulness, which I was keen to retain, and it had to feel modern to reflect the taste of my client and his young family. I knew I wanted the garden to be full of roses, scrambling over the walls and in flower beds in the soft, romantic way they do so famously at the Hall, but I wanted to reflect the freshness of the interior of Bowling Green – its brightly coloured fabrics and contemporary furniture, which contrast with its wonderful old paintings – by giving the garden a crisp, contemporary edge, as well as acknowledging its history.

At Bowling Green, as at the Hall, enormous limestone retaining walls support a flat garden in front of the house. It is as if garden and house are held high on a plinth, from where there are magnificent views over the Dales. One of the Hall's formal terraces (which is also known as the bowling terrace) contains a raised geometric lawn edged in stone and with yew topiary at the corners. At Bowling Green, I wanted to reflect this pattern in a different guise, so instead of raising a lawn, I sank one into the centre of the garden. It is only a slight

BELOW
The 'embossed' lawn at Bowling Green reflects the raised bowling lawn at Haddon Hall. It was my starting point, and set the proportion for the divisions in the rest of the garden.

Newly planted yew topiary is clipped into the same shape as that on the bowling terrace at the Hall, and before long will have reached the same size.

indent, a drop of 100 mm with a metal edge to keep it crisp. It just marks the lawn – like an embossed piece of paper on which you can read the pattern as the light catches the slight change in level – and topiary yews are planted at the corners, in exactly the same way as at the Hall.

This 'embossed' lawn acts as the spine of the garden, setting up the division into thirds of the rest of the space. It is centred on the ancient gates, because the predominant view is outwards from the house, but the terrace door does not align. Looking outwards, to the left are two large square beds, and on the right is a plain lawn for badminton and volleyball, and where the children can kick a ball around without damaging any plants. On both sides the proportions of the central squares are marked with clipped cubes of copper beech, which frame the spaces.

The two square beds contain great ribbons of 'grass borders'. I wanted to create a medieval, wild feeling that was simultaneously modern. At the Hall are 'flowery meads' filled with wild flowers: ox-eye daisies, orchids, fritillaries, yellow rattle and lesser scabious in the long grass. At Bowling Green I have created more of a prairie meadow, with herbaceous grasses and perennials such as *Phlox*, *Veronicastrum*, *Sanguisorba*, *Knautia macedonica* 'Melton Pastels' and *Scabiosa columbaria* subsp. *ochroleuca*, with hollyhocks – billowy and 'meadow-like', but in a contemporary way.

Sinuous limestone-chipping paths cross the square beds at 45-degree angles. From one direction they are invisible, and there is simply meadow, but from another viewpoint they can be seen as 'arrow slits' through the beds. They lead the eye out into the landscape, a little like the arrow slits at the Hall, through which can be glimpsed slices of countryside. Within this 'meadow' planting, copper is an 'undercoat' colour that brings modernity and moodiness, and is picked up by the cubes of copper beech. The dark, plummy tones of *Anthriscus sylvestris* 'Ravenswing', dark-centred poppies and the blackcurrant-and-white *Aquilegia vulgaris* 'William Guiness' contrast dramatically with the lively fireworks of the *Knautia macedonica* 'Melton Pastels' that runs through these beds.

Against the house and the back wall are deep borders filled with perennials in the soft colours

In the summer, with the doors and sash windows open, you can drift easily into and out of the house. It was therefore vital to link the garden very closely with the wonderful interiors, which display my client's passion for the contemporary as well as love of the past.

LEFT AND ABOVE
The paths within the 'meadow' borders lead the eye to the corner of the garden and the stunning views beyond. The house sits in the centre of the back of the walled enclosure, and the garden is laid out in front.

OVERLEAF
On the main facade of the house, I wanted to echo the Hall's abundance of roses. *Rosa* 'Ethel' does the job with vigour and good health.

of old chintz, resembling a hand-blocked fabric, with salvias, nepetas, phlox, erigeron and hardy geraniums. A luxuriance of roses tumbles off the house wall. This reflects the romantic English garden that is part of the 1920s planting at the Hall. The border is very loose and 'countrified', and plants are encouraged to self-seed within it. In fact, even when we are setting the plants out, we place them to give a slight suggestion of the haphazardness of self-seeding. The contemporary grass border also has that feel, because the architectural structure of the grasses is softened by the knautia, scabious and hollyhocks, which also self-seed.

When we started work at Bowling Green, there was no garden to speak of; it had been ploughed up by a tractor and used for growing potatoes. Around the edges, inaccessible to the tractor, were plum and damson trees, which may have originally been trained against the walls but had grown into elegantly arching natural trees in a band that runs down the side of the garden. We grow roses through them, and they're still very productive. Underneath the trees, where it is slightly shady, we planted snowdrops, hellebores, perennial honesty (*Lunaria rediviva*) and foxgloves. A couple of surviving apple trees were also kept for their character and maturity. Meanwhile, at the right-hand edge of the garden were old yew trees, probably an ancient hedge that had not been maintained. We cut them back very hard, and we are now beginning to clip them into topiary.

All the outer walls of the garden are clothed in roses that hang down into the parkland. In winter, approaching visitors see the big terrace walls tumbling with the magnificent hips of roses such as *R. filipes* 'Kiftsgate' and *R.* 'Seagull'. The garden retains the structure of the clipped beech cubes – which hold on to their coppery orange leaves – the formality of the indented lawn and the yew trees clipped in exactly the same way as those at the Hall. Just in front of the building, a very simple rectangular reflecting pool sits in the grass. Seen from the house, it reflects the trees of the park and the old gate piers, and from the garden it gives a mirror image of the formal house, with its fine Georgian windows.

Against the house, the deep borders are almost blowsy, with the soft colours of a faded, hand-blocked chintz.

The pale grey-lilac-pink of *Campanula lactiflora* 'Loddon Anna' (opposite, top), *Lunaria rediviva* (opposite, bottom) and lilac (*Syringa* 'Charles Joly') (below left) is energized by the brighter spots of colour in the mix, such as *Cirsium rivulare* 'Atropurpureum' (left).

OVERLEAF
The scale of the 'meadow' borders gives them the weight they need to feel balanced in the garden and to have an impact in the context of the powerful landscape beyond.

Haddon Hall

PREVIOUS PAGES
The ancient house is softly veiled in pink and cream roses, which are allowed to clamber over the walls.

RIGHT
The Dorothy Vernon bridge crosses the River Wye, with water meadows on either side and views of the Elizabethan terraced garden.

BELOW
English Longhorn cattle graze in the park, just as they would have done in the eighteenth century, when their white horns were prized by manufacturers of buttons, cups and cutlery.

OPPOSITE
A doorway leads through atmospheric ancient walls from the garden into the medieval park.

This extraordinary medieval 'great house', set on a limestone slope above the River Wye near Bakewell in Derbyshire, is one of the most romantic places in England. Its enthralling garden is arranged as a series of terraces, retained by enormous buttressed stone walls at the lowest level, the principal terraces linked by a remarkable early seventeenth-century staircase with beautifully preserved original balustrading.

The earliest parts of the house date from the twelfth century, and it subsequently evolved with each generation until the late seventeenth century, since when little about its appearance has changed. What we experience today is very close to what a visitor of 1697 described as a 'good old house, all built of stone, and behind it is a fine Grove of high trees and good Gardens'. In the eighteenth and nineteenth centuries the house was uninhabited and became an object of wonder, a place of legend for the tourists who visited. It never quite fell into ruin, however, and was meticulously restored to its former glory and turned back into a home in the early twentieth century by the 9th Duke of Rutland, grandfather of my client, the current owner, Lord Edward Manners. Lord Edward's grandmother Lady Kathleen planted the garden, bringing the softness and elegance of the 1920s to the imposing estate and making the garden famous for its roses.

I first came to Haddon to work on a garden for Bowling Green, the former games pavilion that sits in its own grounds above the Hall, and where Lord Edward lives most of the time (see page 154). It was wonderful to study the Hall for this purpose, and when we finished Bowling Green I was thrilled to come back and start work on the main garden.

Rather than a historically correct Elizabethan garden, we wanted to create a garden of today. But it was necessary that it should celebrate its history, too, from the romance of the 1920s and the famous roses all the way back to Elizabethan and medieval times. The ancient structure is beautifully put together: enormously high walls hold up the large, flat spaces of the fountain terrace and above that a formal area that was once a bowling alley, where majestic topiary domes mark the corners of the stone edging.

The decision to remodel the garden was a difficult one. The garden was still very strong, but certain areas had lost their way. It was much-loved,

A bay window in the Long Gallery is set with Bombay glass in a diamond lattice that inspired the motif in the knot garden.

both by the owners and among the general public, and had been cared for with passion and dedication, which meant that it was very difficult to strip off the layers that no longer felt right. The garden had to match the atmosphere of the house, and although the architectural plants and other structure were very successful, certain layers of planting – particularly those that had been added in the 1950s – were not of the right palette. Many plants had been put in by the family, and had particular associations and memories that made it tricky to remove them. We worked very carefully, gently removing these layers as though peeling an onion, until we arrived at a base that felt right.

On the fountain terrace odd diseased crab-apple trees blocked the view. Beds of roses of every colour grew out of sterile-feeling bare soil; the mixture of colours must have been very exciting when they were first planted in the 1950s and 1960s, but it had come to look muddled. We wanted to achieve a more coherent planting, one that pulled house and garden together. We left most of the roses trained on the walls of the house, removing just those that were the wrong colour to leave a palette of pink and cream with the occasional shock of orange or apricot. The borders of roses in bare soil were removed. Trees that blocked views were taken out, and we added a new layer of romance and strengthened the link to history and to the house by wrapping a frame of topiary around the fountain terrace lawn. We wanted the atmosphere here to be naive and relaxed, as though the garden had evolved in the same way as the rambling house, and so we used native deciduous trees – mainly beech and hornbeam – for the topiary, rather than evergreens. There were already topiarized yews on the bowling lawn level above, for greater formality and elegance.

The new topiary sits in a wide band of wild-flower meadow. This 'flowery mead' frames the lawn on two sides, and is studded with species roses, orchids and rare and historic bulbs of the type that would have been planted in the gardens in the sixteenth and seventeenth centuries, such as *Lilium martagon* and *L. canadense*. This is predominantly an Elizabethan garden, and so I imagined how it would have been at that time. Elizabethans were fond of selecting particularly beautiful and unusual plants from the wild, so a delicate orchid, rare crocus or

double buttercup, for example, would have been collected and added to a flowery mead of this type. Native meadows are always appreciated for their simple, natural beauty, but we wanted to make this one more bejewelled, with rarities and topiary, in the same way that the house is bejewelled with beautiful windows, opulent tapestries, fine furniture and exotic rarities.

The next layer of historical reference is the romantic English garden, in which roses tumble over the walls and aubretia and a pretty verbascum grow all over the stonework. A soft froth of this kind has always existed everywhere in the garden, and I particularly wanted these borders to reflect the garden of the 1920s, when Lord Edward's grandmother replanted it. We removed many of the odd little grass paths, widened the borders and laid the main paths with chippings of the same limestone as the house. The borders are now a loose and cottagey mixture of roses and perennials, spilling over the paths, romantically soft and casual.

During our research we found photographs of the garden taken in the 1920s. We wanted to reinstate the handsome benches that are shown in front of the borders, so we made drawings, and it was very rewarding to have the estate carpenters make them anew. This was another important part of collecting the past, bringing back the romantic garden and removing anything 'municipal' with a flavour of the 1970s.

At the foot of the ancient stone wall at the back of the garden, we planted a 'Dyeing Border'. Haddon Hall is famous for its tapestries, so to forge a link between the interior and the garden, this border contains plants that were used for dyeing the silk threads used in the embroidery. In front of it, on a level with the bowling lawn, we planted the cleared rose garden with a knot garden in a pattern that reflects the diamond-paned windows above. The knot is formed by *Lavandula angustifolia* 'Munstead', rosemary (*Rosmarinus officinalis*) and germander (*Teucrium chamaedrys*), a plant the Elizabethans loved. Planted within are sweet violets (*Viola odorata*), *Geum rivale* and *Campanula persicifolia* var. *alba*, all plants that were popular in the seventeenth century, and others that would have been used for medicine, such as *Iris germanica* var. *florentina*, known as orris root.

The knot garden is laid out with period plants and hedging.

LEFT AND OVERLEAF
The Elizabethan staircase, with its original balustrade, links the two principal terraces.

From inside the house, the whole garden is viewed from the silver-grey panelled room known as the Long Gallery, which runs the full length of the fountain terrace. This light, airy room has windows on three sides, including extraordinary bay windows, some of which have leaded lights of 'Bombay' glass, tinged green in places, that bulges and catches the light in a most beautiful way. This room gives a remarkable impression of being in the garden. At one end is a view down to the knot garden, which is designed to be seen from the top of the Dorothy Vernon steps (so called because the sixteenth-century heiress of the Hall used the staircase to slip out and elope with her lover) on leaving the house. Pattern-making is important for more than the knot garden, however; the way the garden is laid out, with terraces, means that it is all to be viewed from above, as pattern. In this way it can be seen that the fountain pond is held in by 'brackets' of copper beech, which I added as a modern touch and a link to the garden at Bowling Green.

The house and garden are open to the public, but Haddon is essentially a family home, and a private passion for plants is at the forefront. We kept the paths narrow, so that visitors must weave their way through the plants in single file. The public is invited in, but the garden still feels very private, intimate and special.

OPPOSITE, TOP
In the knot garden, germander hedges are filled with *Geum rivale* 'Album' and *Campanula persicifolia* var. *alba*, and flanked by stately topiary domes.

OPPOSITE, BOTTOM
The Dyeing Border against the back wall contains plants that would have been used to dye the silks for the tapestries in the Hall, such as mullein (*Verbascum thapsus*), tansy (*Tanacetum vulgare*), cardoon (*Cynara cardunculus*) and weld (*Reseda luteola*) for yellow, woad (*Isatis tinctoria*) for blue and St John's wort (*Hypericum perforatum*) for red.

The frame of native wild flowers forms the perfect soft foil for the native topiary. The impressive white *Campanula latifolia* var. *macrantha* 'Alba' (below) grows prolifically among the native flowers, which include sorrel, ox-eye daisy, foxgloves (*Digitalis purpurea*) and valerian (*Valeriana officinalis*), and plants that would have been used in gardens in the fifteenth and sixteenth centuries, such as the gillyflower (*Dianthus*) that is carved into furniture in the house.

FOLDOUT, ABOVE, RIGHT AND PAGES 186–87
A medieval tapestry of wild flowers and topiary lines two sides of the fountain terrace lawn, with its low squares of clipped copper beech.

OPPOSITE
The limestone-chipping path was kept narrow to enhance the effect of the borders and create a feeling of intimacy.

Essential
Topiary

Topiary is marvellous for creating scale and structure and bringing a sense of tradition and longevity to a garden. I use it in many different ways: formally, to delineate axes and frame buildings and entrances; and informally, to create punctuation and permanent, weighty structure; planted as random groups or singly in a courtyard or meadow; or throughout a garden, as at Allt-y-bela (page 24).

Topiary is most formal when the plants are of uniform shape and size, and set out symmetrically. The double row of *Quercus ilex* topiary in the top photograph on this page is mirrored by a double row on the other side of the lawn. Together they frame the house in a very grand way, and, looking out from the double doors, the eye is caught by an obelisk on each of the axes. The trees are very tall, but, because their stems are smooth and clear, they do not restrict views across the garden. Their height relates to the scale of the building, which is very tall, and helps to anchor it, linking it to the scale of the countryside beyond. In the same garden I've also used copper-beech topiary emerging from a stepped yew hedge, dividing a very long herbaceous border (right, middle). The beeches punctuate the border,

OPPOSITE
Topiary can be used in varying degrees of formality. In a grand garden, a virtue is made of symmetry (top and middle), while a more relaxed arrangement frames a doorway at Cottesbrooke Hall (bottom).

FAR LEFT
Trees in a mixture of sizes and shapes, but all beech, create an anchoring weight to the planting at the corner of a house.

LEFT
Contrasting shapes of clipped beech, symmetrically spaced, add height and rhythm to a yew hedge.

BELOW, LEFT AND RIGHT
Informally placed in a long herb terrace, topiary gives permanent vertical structure, appropriate for this historic property and complemented by oak railings.

in terms both of colour – breaking up what could be a monotony of green, and relating to the pinks and plums in the border – and of form, indicating routes through the garden via gaps in the hedges. At Cottesbrooke Hall in Northamptonshire (page 188, bottom) we framed the doorway to the reflecting pool not with a matching pair of topiarized trees, but in a more dressed-down way, with asymmetrical shapes and odd numbers.

I use many different trees for topiary, and particularly enjoy those that are unexpected. Among evergreen species, I use box, yew and occasionally holly). As for deciduous trees, beech is wonderful for the warmth of its golden leaves in winter, but I also use hornbeam and field maple and even hawthorn. Such topiary feels naive, but it forms a very effective contrast to grandeur, such as the wonderful hawthorn topiary at Hatfield House in Hertfordshire. Conversely, I love cottage-garden topiary, but I feel it is more successful when it is made of box or yew, as a contrast with the characteristic, slightly chaotic mixed planting of flowers and edible plants.

The right shape is important: a pyramid would look wrong at a cottage, but a peacock emerging from a hedge or a tiered wedding-cake shape would be fantastic. I am not keen on kooky shapes, though, and I feel simple beehives, domes, cones on balls and cylinders – whole or in layers – work best.

The topiary at Allt-y-bela (right and opposite) creates division in the garden. There are no internal hedges to compartmentalize the space, but the grouped topiary acts as broken lines, separating one part of the garden from another. I've added more 'fancy' layered beech as sentinels at the entrances to the house, and by the bridge where the drive comes in. I think these shapes guide people around the garden, however subconsciously, by drawing the eye.

Grouping topiary of different shapes, scales and types is a bit like hanging varying pictures above a fireplace: a jumble can work together if it is the right collection, with the various images and shapes playing off one another, producing a whole that is greater than the sum of its parts. With topiary, mixing the right materials and shapes will create a relaxed, but still distinctive and weighty presence.

Topiary is particularly important in the winter, and at Allt-y-bela its scale helps to anchor the building, reducing the apparent height of the tower. There is an art to cutting topiary well, and it exhibits the skill of the gardener, communicating the fact that the garden is cherished.

Holme Hale Hall

PREVIOUS PAGES
Paeonia lactiflora 'Krinkled White' is grown against the base of the potting shed walls to provide cut flowers for the house.

ABOVE
Adding wild-flower seed to the existing pastures has brought a more relaxed, 'country' feel. The ox-eye daisies particularly love the well-drained, sandy soil.

OPPOSITE
Sentinel beehive yew topiary creates more formality on either side of the front door. A narrow spike of water emphasizes this framing, and draws the eye from inside outwards into the garden.

When I first came to this wonderful early nineteenth-century house in Norfolk, it was cut off from its ancient parkland by a huge shelter belt of laurel and holly. The house was hemmed in, but the space within those trees felt very empty, almost barren and cold. A huge copper beech tree had blown down, and I had been asked to give my ideas about what to do in its place. When I saw the walled kitchen garden, however, I was captivated, and I persuaded my client to start the redevelopment of the gardens by concentrating on this area first.

What fascinated me about the walled garden particularly was its location, very close to the front door and the formal facade of the house. Furthermore, the enclosing wall nearest the house sweeps down to a mere 1 metre, and above it are delicate railings, simple but elegant, which are designed to invite interest in the vegetable garden by displaying it to those arriving at the front door. I loved the idea of such pride in the growing of vegetables at a grand house like this, and I felt we must reinstate that link and grow vegetables within view of the front door once more.

Although the brick walls and potting sheds constituted an excellent backdrop, the space within had become somewhat forlorn. A vast area of mown grass – broken up only with a couple of goalposts for the children's games of football – had become almost like ridge and furrow from the wear of the mower tyres, which had run up and down in exactly the same way for perhaps thirty years.

The walled garden covers an area of just over 1 hectare – much too big for simply growing vegetables – so we drew up a plan, keeping the vegetables at the forefront but also making room for a late summer flower garden, a tennis court and a swimming pool. We divided the space into thirds, dedicating the third closest to the house, and visible from the upper windows, to vegetables. Against the walls of that third we made wide borders in which to grow herbaceous perennials to supply plentiful cut flowers for the house. The border was divided across its width with a path made from the local crushed limestone, following the lovely curve of the corner walls. This chipping path with its brick edge runs around the entire kitchen garden, like a picture frame, linking all the areas.

Annuals such as the opium poppy *Papaver somniferum* 'Black Paeony' are grown among the perennials in the outer borders.

Phlomis russeliana, with its pale lemon flowers and silvery-backed heart-shaped leaves, creates a perfect highlight and contrasts with the pale blues and pinks in the border.

Almost the very first thing we did was to build a greenhouse, leaning against the wall where there was evidence of a previous one. The wall swept upwards to a generous height for a single-pitched roof, and when we dug the new foundations we found that we had designed the new structure with almost exactly the same proportions as the original. It is made traditionally, with painted wood and narrow-paned fish-scale glass so that rainwater runs down the centre of the panes, away from the timber. The beautiful central gable gave us a strong centre point with which to line up the path that divides the four new beds. The cross path was lined up with the original gate in the railings, and this axis runs right through to the central part of the garden.

While all the flowers in the vegetable area of the walled garden are in soft purples, whites and lilacs, we call the next third of the space 'the hot garden'; there the colours are more vibrant, with reds, oranges and purples from asters and heleniums that come into flower later in the season. This third is divided from the vegetable garden by two large 'tapestry' hedges of hornbeam and crab apple, which start at the height of the walls and dip towards the centre. Three arched openings in the hedge connect the spaces, but it is high enough to form a complete visual divide, giving each area and different planting palette its own space. This mixed hedge is wonderful in spring, when the crab apple blossoms just as the fresh green of the hornbeam is emerging. Later in the year the crab apples add that 'essence of fruit' so suited to a walled garden. Beds of late-flowering perennials surround a cool green lawn, and at the edge is the single remaining original tree, a gnarled old quince. In the centre of the lawn we found the original dipping pool, which we reinstated as an ornamental pond. To dress down its formality and connect it to its setting, we added bulrushes from the old flight pond for the ducks at the back of the estate farm. The giant sculpted snail shell by Thomason Cudworth is intended to frighten off all the other snails in the garden! The same low geometric box hedging that wraps around the vegetable beds also frames the flower borders here.

Through the second hedge, the final third of the walled garden is where the swimming pool and the tennis court are placed. In this very simple green garden, pleached whitebeam (*Sorbus aria*) brings

The original eighteenth-century railings create a transparent link between the formal facade of the house and the vegetable garden.

Planted against the existing brick walls, deep borders full of traditional herbaceous plants in a palette of soft whites, blues and creams contrast with the dark leaf of the plum tree, *Prunus cerasifera* 'Nigra'.

ABOVE, OPPOSITE AND PAGES 202–3
A new greenhouse was built to sit in the position of the original, where the wall swept upwards to accommodate the extra height.

height and a contrasting silvery leaf, and clipped yew topiary sits in chippings, with ox-eye daisies and *Alchemilla mollis* softening the edges.

All the walls have now been covered in newly trained fruit trees, which took a little while to become established but are now very productive. Some of the original woodwork of the old potting sheds and doors had been replaced, but not sympathetically. We were lucky enough to find old photographs showing how it was originally, and so we remade the carpentry details, putting back lovely lattice frames for the apple storage sheds, which must be properly ventilated.

We kept the planting very simple around the potting sheds, retaining the gnarled old pears but giving them a rejuvenating prune. We also pruned an enormous fig tree very hard. It hadn't fruited for years, but now that it is kept tightly fan-trained, it fruits prolifically. Beneath the fruit trees we planted peonies of just one variety, the single-flowered *Paeonia lactiflora* 'Krinkled White', beautifully crisp with the green of the fig and the pears. In late summer the peonies make way for nerines. Both flowers are traditional in kitchen gardens, and are excellent for cutting and bringing into the house. In early spring there are scented violets, which are perfect for making posies for bedrooms and another evocative plant that would have been seen here in previous centuries.

When we finished the walled garden, we turned to the front of the house, the area I had originally been called in to look at. When the heavy shelter belt of hollies and laurels came out, all of a sudden there was a magical connection between the house and the gently undulating parkland. There were lots of ancient oaks and chestnuts, and we planted more trees in 'roundels', which helped to give them an instant sense of age. Planting trees in groups in this way makes them race up and form a single, bigger canopy, giving the impression of a large, old tree.

At the edge of the drive, we added swathes of wild flowers to bring softness and romance. The drive had been rerouted in Victorian times, when the fashion was to make the approach to a house as long as possible. It meanders along the edge of and through a woodland, which is rather magical, but it delivers you to the side rather than the front of the house. To return importance to the facade,

we used estate railings to open up the long view between the park and the house, and then framed the view with two long borders. Immediately in front of the house we've put back a 'carriage turn', and at the centre of that a shallow pool of water gives focus to the central axis of the facade. The two long borders are perennial grass 'meadows', contemporary and bold in feel, a mixture of *Miscanthus sinensis* 'Kleine Silberspinne', *Veronicastrum virginicum* 'Lavendelturm' and pink and blue chicory to flower later in the summer. In the light sandy soil the chicory does almost too well, seeding itself prolifically about, but it's all contained within the two long borders, formally perpendicular to the house and framing the central sweep of lawn. The symmetry of the facade itself is softened by an ancient *Rosa banksiae* and a fabulously ancient-looking wisteria, which fills the air with incredible scent when it flowers.

Formal apple arches draw the eye to the centre of the walled garden, where the reinstated dipping pool creates a focal point with its sculpted snail shell by Thomason Cudworth.

Parallel low cloud hedges of box edge borders filled with perennial grasses, which frame the formal facade of the house when seen from the park.

The spears of *Veronicastrum virginicum* 'Lavendelturm' are threaded through *Miscanthus sinensis* 'Kleine Silberspinne', whose reddish-silver flower heads, which appear late in the year, turn pale brown and last long into the winter.

South Wood Farm

PREVIOUS PAGES
We divided the garden from the meadows using simple estate railings that have been allowed to rust. Painted, they would have been too formal and crisp, but the orange of the rust knocks them back a little. *Tulipa* 'Jan Reus' is planted sparsely in the grass.

BELOW
Planting old espaliered pear trees immediately gave the garden antiquity. They enclose an 'anteroom' to the main courtyard.

I first went to South Wood Farm, in the rolling Devon countryside, on a magical early summer's day. As the tiny lane became smaller and smaller, I was reminded of my childhood in Dorset, when my godmother would take me on long rambles to find extraordinary gardens and plant nurseries. These explorations were the sort of trips on which we didn't quite know where we would end up, and that's how it felt to arrive at South Wood Farm.

The first thing visible on turning in at the gateway is a beautiful yeoman's farmhouse, with ornate carving and detailing, and stone-mullioned windows at the front and oak mullions at the side and back. Next to the house is an old threshing barn with a rusty corrugated-tin roof. South Wood is an atmospheric classic medieval cross-passage house, with a history of occupation since the time of Henry III (1210). It was a working dairy farm until 1993, when the previous owners bought it and did some minor restoration work on the house, but the buildings and yard were still derelict when my client took it on in 2005. The garden had a lot of charm – it had been gardened, not designed – and it was a great pleasure to eat lunch in the midst of the colour and activity of the main enclosed courtyard, filled with flowers spilling over the paths and busy with bees and butterflies. But there was no proper structure, just a bay tree in the middle and a collection of crisscrossing paths. I wanted to keep the humility of the farmhouse, but the refinement that had existed had been eroded over time, and I felt I should give it back some grandeur, and place the courtyard properly in the hierarchy of the garden once more.

There was no celebration of the front of the house: its inhabitants and visitors alike parked next to the garden wall and entered through a little gate at the side. The first thing we did, therefore, was to change the point of arrival to the house. The trick was to bring the cars right through the old farmyard, which we re-cobbled completely. Rather than parking them there, we created a lean-to cart shed for cars to pull into, behind the old barn. We used the same cobbles to form a little ribbon of path leading from the parking place into what I call the anteroom, an outside 'hallway' – not a complete enclosure, simply separated by espaliered pears on one side. The tracery of the pear branches leaves the

Clipped cubes of yew interplanted with lavender flank the main path to the front door. In the Stone Orchard (bottom), beech squares in a grid around each tree hold on to their golden leaves over winter.

On the south side of the courtyard, where the wall is low, a row of espaliered crab apples (*Malus* 'Evereste'), smothered with blossom in spring, increases the sense of enclosure.

Native spring flowers, including lady's smock (*Cardamine pratensis*; right), cowslips (*Primula veris*; far right), primroses (*Primula vulgaris*) and snake's head fritillary (*Fritillaria meleagris*), stud the sunken path from the formal orchard to the courtyard.

beautiful wall visible, and yet there is the sense of stepping into an enclosure.

The anteroom forms part of the slow arrival, with its layers of suspense and surprise. These spaces are almost a stage set: the narrow lane; the simple country drive with grass in the middle, lined on one side by estate railings, with views of the rolling countryside; the cobbled farmyard and parking area – and still, although the visitor has glimpsed the side, the front of the house hasn't revealed itself. The journey continues on foot, with the tactile experience of walking on cobbles, through wild flowers, into the anteroom. Through the gap in the espaliered pears, with an impressive stone wall behind, carved stone steps lead up to a handcrafted oak gate and into the courtyard, where the beautiful facade of the house is finally seen in full, and a path leads to the front door.

The front of the house has a double-height porch and doorway that were built for show, and naturally the intention was to arrive on an axis with this door. It is not in the centre of the building, however, although the entrance through the garden wall is. In order to be able to take a strong, straight path from the main gate, we created a diamond path around a thyme lawn under the bay tree in the centre of the courtyard garden, then used a path running across it to deliver people on to the main axis to the front door on one side, and to the orchard gate on the other. We flanked this main path with pairs of clipped yew cubes that we were lucky enough to find on a nursery trip to Belgium, and they mark it out as the most important path in the courtyard.

The enclosing walls of the courtyard all still existed, but had lost their refinement. Clunky stone capping, 'cock and hen' style, made it feel too much like a cottage garden, whereas throughout the house the hand of the craftsman can be sensed clearly, in the hewn stone flags and ancient, smooth limecrete on the floors, naive plasterwork on the ceilings, and old wooden doors and panelling. All is simple, elegant vernacular of very good quality. We wanted to echo and celebrate this in the new craftsmanship in the garden, so we changed the capping on the courtyard walls to simple blocks of the same tooled Ham stone as the dressings on the top of the porch, and added a wide, beautifully crafted oak gate. The wooden posts from which it hangs are hand-adzed

From the parking area the journey to the house continues on foot, with the tactile experience of walking on cobbles, through wild flowers, into the 'entrance hall' of the main courtyard.

(cut with an axe-like tool rather than a saw), and the tops are carved with oak leaves. We used the pattern of the banister spindles inside the house as a model for the lattice in the top half of the gate, so the carpenter who built the house has a direct hand in the new gates that lead around the garden today. One of the enclosing walls is low, and since I wanted the house to sit completely in its own space (without the visual presence of the barn), we planted a line of pleached crab apples to increase the sense of enclosure and to relate more closely to the height of the walls on the other three sides.

Inspiration for the design of the courtyard came from early eighteenth-century engravings of little enclosed gardens filled with pattern. Such pattern-making felt right for the courtyard, but it was important that it was not too fussy, and not all filled with flowers. Four trim lawns around the thyme and bay tree give it that simplicity, almost like a very small, dressed-down college quadrangle. In the herbaceous borders it was essential to use topiary, which suits the period of the house, and lavenders, musky plants and old-fashioned roses to fill the air with perfume. Not lots of different roses, however; we wanted to keep it simple, so that it felt as though the yeoman farmer who lived there had obtained a cutting of a special and rare rose, something he had never seen before. Then he produced another cutting and another cutting, and then perhaps he got another rose from a friend. Rather than a lot of varieties, the impression is of a few treasured roses that have been propagated.

This central courtyard, which catches the sun, is the core of the garden, and the principal garden 'room'. The main reception rooms overlook it and the front door of the house is there, as are all the flowers, the detailing and formality, and the beautiful gate. It leads to all other areas of the garden.

Other spaces in the garden were also contained, and they felt magical. We realized that containment gave a feeling of security and protection, and also that it enabled us to create bejewelled gardens full of pattern and productivity that didn't feel at odds with the rural setting. Although the house is surrounded by a generous piece of land, it was very important that it sat in a garden that was not too big, and so early on I made an effort to pull the garden in tight around the house, allowing the outer areas

BELOW
The house is immersed in wild flowers, connecting it instantly to its history.

OPPOSITE
The herbaceous borders are also filled with roses trained on domes, with the spires of native foxgloves (*Digitalis purpurea*), *Aconitum napellus* 'Schneewittchen' and the dark lupin 'Masterpiece'.

The oak gate leads to the kitchen garden, where *Lavandula* × *intermedia* 'Edelweiss' frames the central lawns. In one corner of the courtyard we built an enormous oak bench with a high back, and there is space to seat twelve at the table, to dine immersed in the scent and the colour of the garden.

THIS PAGE
Rosa 'Félicité Parmentier' (right), *Tulipa* 'Apricot Beauty' and the wallflower *Erysimum cheiri* 'Blood Red' (far right) are mixed in front of the lead waterspout set into the courtyard wall.

OPPOSITE
With the back and front doors open, the view through the house ends in a spout that fills a water trough.

such as the farmyard and barnyard to feel more agricultural. I wanted the garden to be like a beautiful piece of embroidery around the house, but one that very carefully unravelled and became simpler and more agricultural as it flowed out and met the landscape.

We knew the garden should not be too ornamental, but rather should be steeped in husbandry and craftsmanship and productivity. The property has a long history of growing fruit, herbs and vegetables, and I felt this vegetable-growing should also take place in an enclosed space. To the side of the house, we enclosed a small area with the house wall and an existing stone wall (which we repaired); the other two sides are hedged rather than walled, to preserve the importance of the main courtyard. We created a mixed tapestry hedge of yew and box (which will be allowed to grow to 2 metres high), the mix slightly dressing down the formality of the plant material. The raised oak vegetable beds are set in a mixture of chipping and grass paths; step-over apples and pears surround them; and four oak fruit cages, like banqueting houses or pavilions, are covered with traditional galvanized chicken wire.

From this vegetable garden a gap in the original stone wall leads to the area that we call the Stone Orchard, where we planted only damsons. We wanted these fruit trees to feel more formal than those planted elsewhere on the property. The orchard is also enclosed, and we reinforced the formality and gave it a modern twist by planting the trees within a grid of clipped squares of green beech. From benches placed at the end of the rows there are beautiful views back to the house. From there a path leads down to the randomly planted Pip Orchard (of apples and pears), which connects to the undulating landscape, allowing the garden to dissolve away.

This organic garden enhances the habitat for wildlife and is planted with species that insects and birds like, such as those that give berries and hips. Without a doubt, maintaining the balance of plants and managing the upkeep of the garden requires a lot of careful work, but my client is a keen gardener, so I always knew it would be gardened with passion.

Tools are stored neatly in the potting shed, clean and ready for use. In the greenhouse there is always a collection of flowers from the garden in small vases. The potatoes on the potting bench are chitting (sprouting) in labelled rows.

In the productive kitchen garden, flowers like these magenta sweet williams (left) are grown for cutting, and are mixed with abundant crops of vegetables. All the plant supports and protective frames are made from hazel, artfully woven into shape by the gardener.

Essential
Craftsmanship

It is a great pleasure to incorporate craftsmanship and husbandry of the land into our gardens, and to allow what we create in our gardens to feel specific to the site. In some woods, such as our local one at Allt-y-bela, timber is still drawn out by horse. Keeping these traditionally worked woodlands going is essential to the gardens we create. My gardener at Allt-y-bela, Steve Lannin, is seen opposite bottom right with hazel rods for making rose domes.

Beneath the layer of plants in every garden is another layer, often more permanent: the built structure of the garden. To create that I rely not only on landscape contractors to shape the earth and build up the levels, but also on craftsmen and -women who make the final surfaces and special details. For me, installing a standard gate or an off-the-shelf bench represents a lost opportunity; instead, I am looking for uniqueness, and for elements that connect a garden to its location.

This ethos is followed through into the plant supports and structures made by the gardener. Although they do not count as permanent architectural detail, plant supports and fences made from local materials help to establish a connection with the surrounding landscape, and with its traditions. We don't use bamboo, but rather (most often) traditional young hazel rods and pea sticks, acquired as locally as possible. Hazel comes from woodlands that are managed by coppicing as they have been for centuries. Every year a coppicer takes out a certain proportion of the straight new rods that have been shooting up from stools that were cut to the ground two or three years previously. The

Alison Crowther works with enormous pieces of timber to create sculptural pieces of great character. Far removed from the bland mass-produced benches that are so often seen, her work gives a real sense of individuality to a garden, as in my design for the Chelsea Flower Show in 2012 (below).

hazel grows as an understorey, beneath the canopy of larger trees like chestnut and oak that will also be carefully managed to supply fencing material and timber for all sorts of other uses, including for gates, benches and tables. For the traditional husbandry of the boundaries we look to hedge-layers, who also use hazel or chestnut for the stakes in the hedge, and to weave a delicate plait through the tops of the posts, like the rim at the top of a basket, which keeps the whole thing in place.

We must use and appreciate these skilled craftsmen in order to keep their trades alive, but we also need craftsmen, sculptors and artists who use these traditional materials in a new way. On a number of projects I have worked with Alison Crowther, who carves massive, bulky pieces of oak into wonderful seats and sculptures. I was fascinated by the way she carves along the medullary rays in the timber, forming richly textured surfaces that bring out the characteristics of the wood. For my garden at the Chelsea Flower Show in 2012 (left), I wanted the contemporary feel of a sculptural bench by Alison; however, it is interesting that, because of the sensitive use she makes of her natural material, although her work is fresh and contemporary it also sits very happily within a more historic environment.

Often the key to bringing elements in the garden together, and to ensuring that they resonate with their surroundings, is the material. Some can be even more site-specific than wood. Stone, for example, varies greatly in England, even over a relatively short distance. Walling made from stone that splits naturally into thin pieces, for instance, will vary greatly from that in an area of hard granite that has to be cut into blocks. We must treasure the wonderful peculiarities of regional variation, because sadly they are being slowly eroded. It is quite as likely that a garden will be paved with Indian limestone as it will with sandstone from the local quarry, but the former lends no intrinsic sense of belonging.

I do not like to 'distress' things falsely. I like the honesty of a newly made gate or bench, but I pay attention to the way in which such an object is crafted; if the right materials are used, the appropriate weathering will take place naturally, and the piece will very soon appear timeless.

All gates and posts are based on architectural details that are found in the house belonging to each garden, or in the local area.

When we made the oak gates shown on this page, the posts were formed with an adze, and the spindles were turned to echo the pattern of the banister spindles in the house. The top cap was carved with oak leaves to reinforce the link to the oak trees growing close by. The latches and hinges were made and hand-beaten by the blacksmith who also weighted the latch to give a satisfying clunk as it drops into place (opposite, bottom right). It was a wonderful collaborative experience among myself and my client and a whole group of specialist craftsmen. These handcrafted elements can become the identifying symbols of a place, and because they are the things you touch, they are a key part of how everyone experiences the garden.

Queen Anne House

PREVIOUS PAGES
Trained apple trees line the paths and every vegetable bed in the walled garden, producing a profusion of blossom in early spring.

RIGHT
The wooded drive gives way to the house in its tapestry of green.

This classically proportioned house in West Sussex has the serenity so common to those that sit next to parish churches. It is approached along a shady wooded drive, from which the house suddenly appears in its sunny glade. The garden immediately around the house is very green and plain, with a few borders close to the building itself. The wider parkland has water meadows and a lake, and the essence of the place is this green palette and the views.

Around the house we simply enhanced what was there by adding to the green texture and replanting the old borders. We did more in the remarkable enclosed kitchen garden of about 0.3 hectares, walled with local stone. There, the most beautiful seventeenth-century dovecot creates a manorial feel, and adds to the sense that this walled garden was for the production of food for a large household. The enclosure is entered through a beautifully forged moon gate – an imaginative addition from the 1930s – with a figure in the centre, holding a hammer and banging on an anvil; it is clearly an image of the blacksmith who created the gate. One of the best things about the walled garden is that all the structure was already there, and the fabric was complete. Central paths edged in natural stone divided the space into four in the traditional manner, and along these and the outer paths were espaliered fruit trees. These ancient, gnarled trees were growing in beds full of ground elder and couch grass, but among all the weeds were self-sown lupins, aquilegia and clouds of *Alchemilla mollis*. The soft, gentle atmosphere reminded me in its impressionistic feel of a painting by Monet.

That first visit persuaded me that the walled garden must be full of flowers. I was convinced, too, that the layout of the vegetable garden must not be disturbed. The four beds were enormous, however – far too big to manage as single beds – so I looked at ways of subdividing each quarter. My client wanted this garden to produce vegetables, but it is much bigger than is required to provide for today's smaller household. Since she also has a passion for cut flowers for the house, we decided on a mixture of vegetables and flowers, keeping the walled garden true to its essence as a productive space.

Three of the newly subdivided quarters contain raised beds with a mixture of vegetables, flowers

OPPOSITE
Behind the house, a sculpted box hedge and topiary yew balls form a green plinth.

PREVIOUS PAGES
Outside the moon gate that leads into the walled kitchen garden is an early summer border of alliums, white *Anthericum liliago* 'Major', foxgloves and bright pink dianthus.

LEFT
The borders that line the outer paths of the walled garden are thickly planted with the tulips 'Florosa', 'Burgundy', 'Queen of Night' and 'Recreado' for the spring.

for cutting and soft fruit. Raspberries, blackcurrants and gooseberries grow as bushes, and parallel bars slotted into oak posts support fan-trained cherries and raspberries and thornless blackberries. At the end of the garden are two greenhouses, one for propagating and growing early vegetables and tomatoes, cucumbers and peppers, and the other for overwintering plants and propagating flowers. The fourth quarter is now a picnic area, with a double spiral path dropping down into a sunken lawn. The spiral walkway is entered through a series of serpentine grass paths etched into the ground. I wanted it to feel ancient, so we 'dished' the grass to give the impression that it had been trodden down over the years. The paths meander through borders of loose prairie-style planting and herbaceous grasses, with tall topiarized copper beeches. The house and dovecote are large and the garden walls high, so it was very important that the planting throughout the walled garden should have an appropriate sense of scale.

While the picnic quarter is loosely planted, its paths flowing, the rest of the garden is tightly organized, with perfectly straight beds and paths of limestone chippings and grass to provide a break from the stoniness of the surrounding walls. For the same reason, the new subdivisions within the existing beds are edged with a combination of oak, steel and brick. The new labyrinth of beds is still accessible and easily managed from smaller paths, helping with the maintenance of the garden and avoiding the need to walk over too much soil. Box hedging strengthens the new subdivisions, and is cut with straight sides but wavy tops, which catch the light and lead the eye along the new lines. All the beds of cut flowers run in one direction in raised oak planters; the vegetable beds run the other way and are further defined by step-over apple trees, which surround them.

The outer reaches of the walled garden consist of borders planted with herbaceous perennials, with a stunning layer of spring bulbs. I wanted to create two distinct seasons within the perennial planting. The first is an exuberant flowering in May and June, starting with apple blossom and, beneath the fruit trees, peony 'Molly the Witch' (*Paeonia mlokosewitschii*), *Geum* 'Lemon Drops' and lots of *Veronica gentianoides* providing a contrast between

The tall viridiflora tulip 'Florosa' – with its soft pink tips that gradually mature to a deep rose, over a creamy white base with fresh lime-green flames – carries the show in this combination. It is a late, long-flowering variety. Behind the tulips the low structures of woven hazel are ready to support perennials later in the year.

OVERLEAF
A view across the walled garden towards the house shows the combination of vegetable beds, edged with brick and step-over apple trees, and cut-flower beds edged with oak.

BELOW
Fruit trees and roses bring height to the beds either side of the path that leads to the greenhouses.

OPPOSITE
The soft, romantic palette of roses, daisies and geraniums sits happily with the old grey walls of the garden.

Masses of peonies, including the dark pink *Paeonia lactiflora* 'Kansas' and the pale pink 'Sarah Bernhardt', are grown for cutting for the house, and hazel wigwams are prepared for sweet peas in the cutting beds, and for climbing beans and peas in the vegetable beds.

sulphur yellow and pale blue. As the summer progresses there are roses, nepetas, foxgloves and astrantias, and many white lupins. The lupins that were already in the garden have been replaced with the very subtle creamy-white *Lupinus* 'Noble Maiden'. Then the garden settles down for the summer, and the gardener keeps the borders full of various annuals. The second major flourish of flowers starts in late August and continues until October, with a variety of autumn-flowering asters and *Gaura lindheimeri* and Japanese anemones.

There are two principal exits from the walled garden: the moon gate, on the other side of which is an early summer border; and through the potting shed, beyond which is a serene pool garden. A lot of work was done to the house and garden in the Edwardian period, including all the stonework in this little sunken garden. It has an ancient yew hedge to one side, and we kept the border in its original place and filled it with similar flowers to those in the kitchen garden, for consistency.

One of the most attractive parts of the garden is a wide terrace at the back of the house, planted with a huge *Magnolia* × *soulangeana* and a wisteria and softly carpeted with creeping thymes and *Alchemilla mollis*. To keep the view simple, we put in a cloud hedge of box, forming a green plinth for the house to sit on. Views from inside the house look out over this flowing hedge. On the upper level, against the terrace, the lawn beneath a double row of very simple yew balls is studded in spring with the early *Crocus tommasinianus*.

For me as a designer, the remarkable thing about this garden was to be able to keep the planting around the house soft and green and simple, knowing that the jewel in the crown is through the moon gate in the walled garden. There, almost all year round, is an explosion of growth and productivity, and even in winter, when there are few flowers, there is the structure of the apple trees. Despite its size, the enclosure feels almost intimate, and it is devoted to the art of gardening. This demanding garden requires plants to be continually divided and propagated; it is a celebration of real horticultural skill, and the time and love that are lavished on it pay enormous dividends in the glorious show.

Raspberries and currants grow against the bars held by a grid of oak fruit-training posts.

The productive quarters of the kitchen garden are set with grass and chipping paths in a pattern. The enormous copper-beech topiary brings scale to the fourth quarter, with its prairie-style grass borders and sunken picnic lawn.

Rows of potatoes are neatly planted in the apple-edged beds. Behind and to the side are the replanted box hedges, which reinforce the divisions within the walled garden.

The outer borders are filled with roses and perennials, and provide even more flowers for cutting.

A blackboard hung by the door of the potting shed lets everyone know what is ready for harvesting. Collections of potted auriculas are kept ready for displaying on a table on the terrace or in the house.

BELOW
The border around the sunken pool garden reflects the planting in the kitchen garden of late summer flowers, such as asters, in a mix of pinks and purples.

OPPOSITE
The Edwardian sunken pool between the potting shed and the house is anchored by four clipped box cubes. Its paving is colonized by *Alchemilla mollis*.

LEFT
In autumn, the kitchen garden takes on a golden hue, with deciduous leaves contrasting beautifully with the neatly clipped box hedges that divide the garden into quarters.

BELOW
The yew balls and sculpted box hedges are crisply clipped for winter, to offer interest when all the flowers are gone.

Guanock House

PREVIOUS PAGES
Lavandula angustifolia flanks the path under the huge yew topiary, filling the air with scent as you brush past.

RIGHT
The garden was a wonderful place for entertaining and performances. Here, the actors of Punchdrunk theatre company sit among our guests in the lime walk before their marvellous 'roving' performance of *A Midsummer Night's Dream*.

When we moved to Guanock House in Lincolnshire, between Peterborough and the Wash, in 1990, gardening was my burning passion. I had been working with Peter Hone at Clifton Nurseries in Little Venice, London, buying and selling remarkable garden antiques. I was becoming more and more interested in formal gardens, and I had become obsessed with the engravings of Johannes Kip and Leonard Knyff, who in 1708 published the *Britannnia Illustrata*, showing the principal seats of the nobility and gentry of Great Britain. I pored over the details of those gardens, which were shown in wonderful bird's-eye view, and absorbed the pattern-making, the detailing, the architectural style. I visited the Dutch Baroque garden at Het Loo Palace in Apeldoorn, built as a 'pleasure house' for William and Mary in the 1680s, and it blew me away. It was a Knyff and Kip illustration brought to life. Its green tunnels and grandly scaled pattern-making fuelled my passion for this period, and I became intensely excited about creating a garden of my own.

Guanock House, which is in a part of the Fens called South Holland, felt somewhat Dutch but also naive, and was of course on a much more modest scale than any of the properties in the engravings I had studied. I wanted to incorporate architectural elements inspired by those engravings, but interpreted in a naive English manner. The house was quite formal in its symmetry, but the detailing was not grand. It had very simple tumbled-in gables, rather than the elaborately patterned gables that are common in Suffolk and Norfolk. The large windows and the big, centrally positioned chimneys gave grandeur to its simplicity. The garden I created was quite grand and formal, but I kept paring it back, knocking the edges off, so that it felt like a piece of old furniture – not over-polished, perhaps with water marks, richly patinated with its history.

I kept an analogy in mind: Sir Marmaduke Darrell, who built the house in 1699 (and probably used it as a retreat, where he could come and shoot duck), also had a house in Holborn, London. I imagine that in London he would have dined in grander houses than his own, where he would have seen wonderful things like a silver sugar shaker, beautifully made and topped with a pineapple, a Georgian symbol of hospitality. Back in the country,

At the heart of the garden, an ancient yew is perfectly in line with the centre of the house. We began clipping it back hard, and by the time we left Guanock it was a dense and flawless topiary drum.

ABOVE
In line with the front door and the yew tree, an opening in the boundary hedge keeps a strong connection, even in the formal part of the garden, with the landscape beyond.

OPPOSITE
To one side of the house I kept the garden completely open to the landscape. I loved the way the building appeared to rise out of a sea of winter wheat.

wanting to replicate such a thing for himself, he might have asked a craftsman to make him a sugar shaker, because he felt it was a smart thing to do, and a fitting celebration of his expensive sugar. But he could not afford silver, so it was made of pewter (poor man's silver); the local craftsman, who had never seen a pineapple except in Sir Marmaduke's sketch, might have thought it looked more like a pine cone. So the sugar shaker for Guanock House ended up beautifully made, but still a pewter sugar shaker with a pine cone on the top. It was this dressed-down, naive impression that I wanted to make with the garden.

The house sits on a tiny rise in the flat, arable land. Before the Fens were drained, several centuries ago, all this was marshland, and at the bottom of the property a river known as Lady Nun's Old Eau once ran. An avenue of walnut trees led up to the house from a bridge over this large, navigable river, which flowed up through Norfolk via King's Lynn, and was part of a very important pilgrimage route to the religious centre at Walsingham. Before the new house was built in 1699, Guanock, an old local word for 'hostelry' or 'resting place', was a monastic grange, and it would have been a place for pilgrims to stay. I wanted to keep that sense of hospitality – in the garden as well as the house – and make a retreat from London for William and me, and a special place for entertaining friends.

We kept the house as pared back and as original as possible, peeling back paint and paper on the interior walls to reveal the layers of history. We wanted that feeling of layered history in the garden, too, but had only a flat field to work with. As I began the garden, I faced something of a dilemma: the open agricultural landscape had a certain magnificence, and I could have allowed the cornfields to come right up to the house, but I wanted to create a real garden, and to do that in a place like this I had to make some sort of enclosure. I wanted my garden to feel like a place of safety in the vast landscape, but also to stay connected to the outside, so I made openings in the hedges to allow glimpses out, and on one side I kept the house entirely open to the fields.

With the house as the centre point I began to compartmentalize the garden by building walls and planting hedges. It felt natural to frame the house's

By mowing a pattern into this big area of grass we turned it into a simple parterre, giving it scale and interest and arresting the eye without being overcomplicated.

symmetrical facade with walls aligned with the gable ends. In the heat of the summer, when the grass was parched, the footings of old walls were visible exactly where I had planned to build the new ones, which gave me great confidence that this was the right thing to do. But the most extraordinary thing about the symmetry of the house and its relationship to the garden was the old yew, planted absolutely central to the front. It was ragged and thin, and I decided to turn it into topiary, clipping it into a strong drum shape. It became the heart of the garden.

Using the new walls and hedges, I created 'garden rooms', and within these new enclosures I was able to create a rich and intricate formal garden. There was a winter knot of copper beech (a less formal choice of plant than ordinary beech or yew), which glowed in the sun; a rose garden with a serpentine box hedge and a pair of pavilions, which separated it from a physic garden; and a herbaceous garden, where I experimented with colour and form, changing the planting completely several times during the fifteen years I was there. To act as a place of transition and dissolve the garden back into the fields beyond, I made a pleached lime walk, which formed a green chamber. I set the limes in limestone chippings in a formal double row around a raised lawn, but I did not want them to be as fancy or refined as those at Hampton Court Palace, for example, which are very precise and symmetrical, with evenly spaced branches. I wanted them to be beautifully trained, but more naively, as a farmer might have done. With this in mind, I was driving past the war memorial in the Norfolk village of Garboldisham when I noticed trees trained in exactly that way, with all the strands tied in, creating a nest-like thicket of branches. It looked magical, and confirmed to me that this was right for Guanock, too.

Between the physic garden and the orchard was a yew hedge, but it would have felt like a Gloucestershire manor-house garden if all the hedges had been of yew. The other internal hedges were more informal 'tapestry' hedges of *Buxus sempervirens* 'Handsworthensis' mixed with either beech or yew. The outer hedges were of field maple or hawthorn, and in the lime walk, an opening to the field beyond was cut through the boundary hedge. I loved the contrasting view from the herbaceous

In winter the agricultural landscape seemed particularly vast and could feel austere; but the garden, cocooned by hedges, in turn cocooned the house.

I loved the perpetually changing nature of the field next to the house, and the feeling of rejuvenation when it had just been ploughed – much like the satisfaction of a freshly dug vegetable bed.

OVERLEAF
The garden at Guanock felt magical, an intricately worked canvas against the vast open landscape, especially when the morning sun of spring caught emerging leaves and hundreds of tulips.

LEFT
The pavilions that divided the rose garden from the physic garden were fabulous for enjoying the warmth of the sun without the chill of the Fens' ever-present wind. One of the buildings had a wood-burning stove, and it was particularly pleasant to light the fire at the beginning of the day and return to have breakfast or lunch there.

BELOW
Clipped copper beech obelisks punctuated the planting of a mixture of pink roses trained on domes and tripods, and the lavender-blue *Geranium* 'Gerwat' (Rozanne), with hesperis and white foxgloves, in the rose garden.

In the long border, *Knautia macedonica* and *Salvia sclarea* var. *turkistanica* are set against the silver leaves of the colossal onopordums. This garden was divided by clipped yew buttresses and Irish yews, which gave it excellent structure in the winter, too. In the facing border I used the pink willow herb, *Chamaenerium angustifolium* 'Stahl Rose' (opposite). It spreads rather, but is a lovely addition if you have the space and can allow it to run.

garden, down the lime walk, through this opening to the vast landscape, and the juxtaposition of mass agriculture with the intensity of my garden. At the side of the house, however, the garden was connected more purely to the fields beyond, with no boundary of any description.

The kitchen garden, by contrast, was completely enclosed. It was in an old fold yard away from the house, to the north-east, backed by agricultural buildings; some of the walls were intact, but much needed rebuilding. It was amazing to have that protected space in the Fens, where it is said that the wind comes straight from Siberia. The walls created a microclimate, which meant that I could bring my vegetables on two or three weeks earlier than our neighbours. I kept the kitchen garden simple and informal, with intersecting grass paths and four vegetable beds to allow crop rotation. In a narrow bed beneath the surrounding walls I put beautiful flowering plants for cutting. On chilly days, it was magical to be able to walk from the house across the grass parterre, step into the kitchen garden, close the door and work in such a protected environment.

Opening the door to the kitchen garden allowed views down one of the axes of the garden, through the lime walk, into the physic garden. The whole garden was built on axes like this; another went through the rose garden to the winter knot, and perpendicular to it was another that, when the gates were open, pulled the eye out of the garden to take in the long view across the fields. An orchard of apples and cherries set in a sea of wild flowers next to an enclosed sheep pasture gave the west side of the garden the softness it needed. At Guanock, some parts of the garden were more gilded than others, but all was in contrast to the agricultural ocean around. When we left, I decided to leave the garden as complete as possible, with all its benches and sculpture, as well as all the plants. Gardens are never static, of course, and I have not been back to see it since we left, but I'm sure the new owners will have gently added their mark, and that the garden is in a great new chapter of its life.

In the spring, the walled kitchen garden was much warmer than everywhere else, meaning I could get on and garden much earlier in the season than my neighbours.

The view through the open door of the kitchen garden led all the way to the fountain in the physic garden, but when the door was closed, the walled enclosure felt completely secret.

Details such as oak labels and traditional protective chicken-wire cloches are important; they are all part of the husbandry of a kitchen garden.

A flamboyant mix of *Primula vialii* and *Iris pallida* 'Aurea Variegata' surrounds the fountain in the physic garden in June. Earlier in the season (opposite), while the plants are still emerging from bare earth, the reflective water is a treat for the eye.

Essential

Pleached, Pollarded and Trained

Pleached, pollarded and trained trees create architecture within gardens. I use them to screen and frame, to line axes and walks, and to form *allées*, divisions and enclosures. The character of the form and the choice of tree create very different effects, but all bring structure. These forms are all created from trees that are pruned regularly to maintain a particular shape and size, and so they are extremely useful in bringing strength and presence to a garden in which space is restricted and a full-sized tree simply would not work.

Pleached trees can either form a semi-transparent veil or act as a solid hedge on stilts, but their height and linear nature (they are literally 'plashed' or interlaced to form a line, be it curved or straight) make them a striking way of creating vertical structure. I love the traditional feel of pleached limes or hornbeam, and sometimes that is exactly what is needed, but I tend to use trees that you might not expect to be pleached, such as field maple or Turkish hazel, English oaks, crab apple or even dessert or culinary apples and pears. Although many evergreen trees can be pleached, I generally prefer to use deciduous species for their more naive effect.

The very different characteristics of pleached and pollarded trees: pleached copper beech frames the path at my garden for the Chelsea Flower Show in 2012 (opposite, top); a double row of pleached Turkish hazel, with their corky grey trunks, forms a colonnade through wild flowers, leading to the gatehouse at Stiffkey Old Hall in Norfolk (opposite, middle and bottom, and left, bottom); limes at Guanock House and flowering pleached crab apple (*Malus* 'Evereste') at Allt-y-bela (left, top and middle).

I frequently use pleached trees in town gardens, where walls or hedges are not high enough to provide privacy from the neighbours' upper windows. Imagine a line of pleached edible pear trees in a town garden where there is very little space against a boundary or against a division within the garden. It brings blossom and fruit, and the bare winter branches allow light into the garden. The branches develop great character, with knuckles and fruit spurs created by regular pruning, forming wonderfully atmospheric green architecture.

In some instances, a more substantial presence is needed, and for this I like to weave the branches of pleached limes or Turkish hazels into a dense, flat, twiggy network that is visible in the winter, and which becomes a solid leafy screen in the summer. To add still more weight I use beech, hornbeam or field maple, and shear the sides as for a hedge or topiary. I particularly love using copper beech in this way, as I did in the garden I created for Laurent-Perrier at the Chelsea Flower Show in 2012. The wonderful thing about beech is that when it is regularly cut it hangs on to its beautiful golden autumn leaves all winter, bringing warmth to the garden on even the coldest and wettest of days.

I also use pollarded trees (those that are cut back to the desired trunk height every three to five years) as architectural elements in the garden. They are more casual than pleached trees, but they can still be used in a formal way. I plant them as *allées* where there is no room for an avenue of full-sized trees, or I use them closer to the house to give vertical structure to a courtyard and frame the building. I love the symmetry and rhythm that this brings to the garden, and the formality is somehow pleasingly contradictory to their rugged, gnarly trunks and the knuckle tops they have just after being cut. We used pollarded limes to create just that formal effect in front of Blackland House in Wiltshire (right, bottom), but if we were using them more informally – on the edge of a river, for instance – I'd choose to pollard willows or alders.

Trained fruit has been a passion of mine since childhood, when I first trained my mother's apple trees 'en arc'. It really shows the hand of the gardener and demonstrates horticultural skill, creating magic by displaying the great care that is taken over the garden.

RIGHT, TOP
A double row of pollarded limes creates a small *allée* leading the eye to an avenue of plane trees beyond the ha-ha at Cottesbrooke Hall in Northamptonshire.

RIGHT
At Blackland House in Wiltshire, the entrance courtyard is enclosed on three sides by pollarded limes.

ABOVE AND LEFT
At Allt-y-bela the pear arches trained on bent hazel rods form a productive entrance to the kitchen garden, and 'step-over' apples bring another layer to the pattern of the beds.

FAR LEFT AND MIDDLE
Espaliers and 'Belgium lattice', which can be trained on a wall or free-standing, create transparent divisions and bring detail, as well, of course, as providing wonderful fruit.

La Frateria

PREVIOUS PAGES
The new pool was kept as simple as possible, surrounded with just trim grass and olive trees.

RIGHT
The pool and house are now strongly connected, and although some rows of olive trees were removed to allow this, the link with agriculture has been playfully retained through the crisp strip of barley in the centre.

La Frateria is perched on the top of a hill in Umbria, Italy, with commanding 360-degree views of the surrounding hills, and a large lake on one side. The house sits in the middle of a commercial olive grove, with a plum orchard on the north side. When I first saw the site, the house was surrounded by trees in very rigid, strict lines, growing out of grass, as is the way with agricultural crops in this area. There was only a tiny area of garden, which was almost circular, like a bastion. At its widest point an area of flat grass diminished and curved away, and in many places it simply dropped off the edge. The impression was of being perched on a high precipice, and that the tiny garden was tumbling off the hill.

The most important aspect of this design was to give the house breathing space, but I also wanted to work with the olive plantation on one side and the plum trees on the other, and embrace the language of this traditional agricultural planting. In order to create space for the garden and address the unmanageable slopes, we cleared four rows of olive trees all the way down through the grove. There were no steps, and the former swimming pool was at the very bottom of the olive grove, far from the house and invisible from it. I wanted to create a wide axis between the house and a new pool closer to it, while also opening up views of the landscape beyond. In this new open space, between the pool and the house, lay the opportunity to design a real garden.

It was crucial to take note of what was around and to use local vernacular details for the hard landscaping. I wished to repeat the many drystone retaining walls that hold up olive groves and vineyards in the area, but instead of rough stone, I wanted crisp, contemporary walls made from precisely tooled square blocks, in a warm grey.

When I began work on the site there was no sense of proportion between the house and the garden, which was simply a steep bank of trees. The new walls created new levels and rectangular terraces that wrap around the house and link it proportionally to its surroundings. As well as the stone walls, I used crisp grass banks and steps to manage the descent to the olive grove and to extend the garden, creating suitable places for seating areas and planting.

The silvery grey-green leaf of the olive dominates the colour palette on the south side of the site. Everything we planted harmonizes with it or uses it as a foil.

OVERLEAF
Crisp grass banks and steps work with drystone walls to create flat terraces that lead gently down to the original levels of the olive grove.

The beautifully scented *Robinia × margaretta* 'Pink Cascade' is planted among the plum trees on the north side of the garden, and also appears as a striking contrast to a copper-beech hedge around a badminton lawn.

Seating areas are shaded by table-top canopies of plane trees (*Platanus* × *acerifolia*) set in a surface of crushed pumice.

It was very important that the colour palette of the plants should harmonize with the setting. The area to the north, where the plums are, is very green, and to the south silvery grey dominates in the olive leaves, but there are also the greens of oaks and elms. We added plants that keep their leaves in winter, such as evergreen oaks, and for striking contrast and modernity we added copper-beech topiary close to the house. The colour combinations are particularly lovely in autumn when the warm russets of the oak leaves appear against the silver of the olives. This mix of silver, green and autumnal russet reflects the natural landscape beyond, with its forests of deciduous and evergreen oaks, and acacias.

The quality of light is striking here, and that affects the way colours appear. Being high on a hill, the garden gets very misty light early in the morning, and glows with the pink of sunset later on. It is in the sun all day, so it was essential to create shade. On one side, where there are grass banks and formal, rectilinear terraces created by the zigzag of drystone walls, the borders are full of greys: lavender, salvia and rosemary, pale blue irises, oregano, agapanthus and many artichokes for their giant silver leaves in summer. (Although there are exceptions, silver foliage has generally adapted to reflect the light, and so these plants do well in hot, dry situations.) Fig trees and ancient vines with characterfully gnarled stems offer shade; and the vines echo those that can be seen on the hillside opposite.

In the countryside near by, I noticed field maples draped with vines planted at the edges of local fields of barley. Since I wanted to connect the garden more deliberately to the agricultural land that surrounds it, to ensure that it felt as though it belonged in this region, I decided to bring this detail into the garden, and between the house and the pool we added a crisp slice of 'field' filled with barley, and pollarded field maples through which grow vines. It is both harmonious and surprising to see these commercial agricultural plants used in a different way. On each side of the barley 'field' is a pergola of chestnut posts, with alternating vines and wisteria forming a roof to shade the walk down to the swimming pool.

We worked closely with the client and the architect to create a pure, modern swimming pool area framed and held in by a simple drystone wall, and we kept the planting by the pool to a minimum.

ABOVE
Sharp zigzag walls hold a planted lower walk that forms the perimeter of the formal garden. Below them the planting is softer, and paths wind gently through the olive grove.

OPPOSITE
Cypresses (*Cupressus sempervirens* 'Pyramidalis'), with their iconic columnar shapes, define the edge of one of the rectilinear terraces, next to the guest house. Rosemary, *Verbena bonariensis*, box and lavender line the path on the narrow terrace below.

Grass comes right up to the simple stone edge, and in the grass are just olive trees, maintaining a cool, tranquil environment. The land drops off very steeply on one side, so we built a large enclosing grass mound to hide the slope from the pool.

At the edge of the garden are the lower walks, the 'dry walks', laid with crushed pumice stone out of which grows a collection of box balls, rosemary (*Rosmarinus officinalis* var. *angustissimus* 'Corsican Blue', *R. officinalis* 'Majorca Pink' and *R. officinalis* Prostratus Group), sage (*Salvia lavandulifolia* and *S. candelabrum*), thyme, oregano and *Calamintha nepeta* subsp. *nepeta* 'Blue Cloud'. The volcanic pumice is local, so it feels right here; the colour is suitable, and it is an ideal growing medium because it absorbs and retains water. These raised walks lead around the edge of the entire garden, almost as though it is a citadel, a walled city. Beyond this last retaining wall, which holds up a raised walk, is the olive grove. Where the olive grove comes close to the garden, we domesticated it slightly by adding serpentine paths and crisscross hedges of bright pink rock rose (*Cistus*). The garden itself is very square and regular, but I wanted the agricultural olive grove to feel more casual, so we kept the paths natural, simply cut into the surface. Among the olives we sowed wild flowers – chicory, oregano and wild Florentine irises – and artichokes, to produce a magnificent meadow in spring and summer. The pale blue chicory is stunning against the olives.

Up by the house, away from the grove, four antique olive trees are planted in a row; they will eventually be tamed and clipped into big topiary drums. Planted underneath is a mixture of thyme, agapanthus and saffron crocus. The nearby town is famous for growing saffron, so we wanted to include it in the garden scheme.

In the main part of the garden, three seating areas are shaded by large plane trees (*Platanus* × *acerifolia*). The surface underneath is again crushed pumice, and the planting is very simple, with pots of agapanthus and scented geraniums.

In contrast to the subtle colour palette in the rest of the garden, the kitchen garden is rich and vibrant. Rectilinear beds of cut flowers and vegetables are crisscrossed with zigzag paths. The beds that run lengthways from the house are edged with lavender, and those that cross the facade with box. From the

Crisp stone walls and angular grass banks form a graphic pattern of terraces. Fig trees give height and provide shade on the terraces, and at the top four ancient olive trees will eventually be clipped into topiary drums.

The dahlias 'Who Dun It' (top) and 'Ferncliff Illusion' (above) are grown with the white *Cosmos bipinnatus* 'Purity' to provide cut flowers for the house. Pots of soft blue-green *Echeveria elegans* perfectly match the candlesticks and painted wood behind.

ABOVE
The striking copper-beech topiary brings modernity to the planting around the house. The clipped shape contrasts beautifully with the naturally lumpy profile of the slow-growing hornbeam *Carpinus betulus* 'Columnaris' behind.

OPPOSITE, TOP
A lawn of *Thymus vulgaris* grows below the beech topiary.

OPPOSITE, BOTTOM
A network of crisp steel rills flow from a smooth cast-concrete pool.

house, layers of flowers in their parallel beds are visible, and the vegetables run the other way, so the view is mostly of flowers: bright pink and purple dahlias; zinnias, nasturtiums and opium poppies in blacks and pinks; and bright blue cornflowers. There are also salad vegetables, tomatoes, peppers, courgettes, herbs and sorrel.

In this part of the garden we again worked with the architect and the client to create a modern, minimalist greenhouse. The downside of such an elevated site is that it can be very windy, and although it is extremely hot in summer, it can be cold in winter. This affects the crops: olives are hardy, but citrus trees cannot be kept outside, so in winter all the lemon trees and the scented pelargoniums come into the greenhouse. In the heat of summer the greenhouse tends to be kept empty, becoming for a time simply a space for a collection of pots, watering cans, architectural details and a few tough plants.

On the north side of the site, overlooking the lake, we kept the design very simple. The plum orchard comes up to new grassed earthworks wrapped with sinuous walls and a curving driveway. Here it is shady and green, a sanctuary from the heat. The mixture of stone and grass banks is the same as that on the south side, but instead of creating formal rectilinear spaces, as there, they delineate fluid shapes that are coolly sophisticated. We planted walnut, apricot and peach trees with the plums, and a fabulous pink *Robinia* × *margaretta* 'Pink Cascade', which smells like wisteria. This *Robinia* is dotted through the plums and also appears on the badminton lawn at the other side of the house. This lawn is approached by a series of big grass banks and steps, and enclosed by copper-beech hedging; the *Robinias* grow out of long rectilinear beds of pumice, with clipped balls of lavender. These beds surround a flat lawn poised above the olive grove, giving spectacular views over the landscape.

Near the house, an intricate garden laid out in a grid provides space for vegetables and cut flowers.

Next to the elegant, modern greenhouse is a profusion of richly coloured dahlias and cosmos, in vibrant contrast to the subtle palette elsewhere in the garden.

The terraces formed by the stone walls and grass banks have given the opportunity for intimate seating areas. Growing out of the chippings are clipped box balls and hedges of olive (*Olea europaea* 'Cipressino').

Against the tall stone walls of the entrance courtyard, the bright blue pool is a modern touch, matched by the simple clipped shapes of the hornbeam (*Carpinus betulus*) and box. Oregano is planted in the chippings, along with *Cistus albidus* and *Verbena bonariensis*, for softness and colour.

It is traditional in hot countries to include the cooling sound of water splashing into a courtyard pool. We worked closely with the client and the architect to incorporate this in a very contemporary way.

OVERLEAF
In the olive grove below the formal terraces, a serpentine path winds through crisscrossing hedges of grey-leaved *Cistus albidus*. In summer these hedges sit in a wild-flower meadow and are smothered in bright pink flowers.

ARNE MAYNARD is a celebrated garden designer known for his large country gardens in Great Britain, Europe and the United States. He draws inspiration from a wide range of disciplines, from architecture to garden history and from interior design to traditional crafts and techniques. Most importantly, he is a passionate and experimental gardener, and his love of plants is evident in all his garden designs. Arne prefers to add structure to a garden through the use of plants and trees, ensuring that the garden can be enjoyed all year round. And there are a number of signature Arne Maynard 'essentials', including roses, kitchen gardens, topiary and loose summery borders, that are drawn out and explored in this book. The garden he is creating at Allt-y-bela, from which he teaches a range of garden courses, is both a showcase for his great gardening passions and a garden in which to trial new plants and techniques. He is the author of two other books, *Gardens with Atmosphere* (2001) and *Garden Design Details* (2004).

WILLIAM COLLINSON has photographed Arne's gardens from the very start. They collaborated on the creation of their gardens at Guanock and at Allt-y-bela, William being a wonderful sounding board for Arne and, through the lens of his camera, allowing Arne to view and assess the progress of the gardens with more than his own eyes. This close relationship has enabled William to capture all the gardens in this book in a way that reflects how Arne himself sees them, providing a rare insight into the designer's own view. William's particular love is, with his favourite camera, to capture atmosphere, especially in landscapes and gardens, where his Hasselblad allows him to render what he frames in great detail, even in large open shots. Photography is now the main focus of William's professional life, and the collection of work you see in this book is a fitting start to a career that has been waiting in the wings since his schooldays, when his passion for photography began.

Picture Credits

l = left, r = right, t = top, b = bottom, c = centre

Most of the photographs in this book were taken by William Collinson. I am also very grateful to have been able to use additional photographs from the following people, but particularly Britt Willoughby Dyer, who has contributed so much to the library of photographs of my work, and to this book.

Rosie Atkins
256

Nicola Browne
192–93, 194, 196t, 201

Nick Hand
56cr

Jason Ingram
134, 140

Rowan Isaac
53bl, 57tr, 200

Arne Maynard
298b

Kristy Ramage
12, 13 (all), 230br, 231tl and tr

Howard Sooley
22t

Britt Willoughby Dyer
14, 15 (both), 16 (both), 17, 22b, 23, 53br, 54 (all), 55 (all), 56tl, 57tl, 57bl, 62–67 (all), 68t, 72–75, 98–99, 101br, 102bl and br, 132–33, 136–37 (both), 141–45 (all), 148bl, 150–51 (all), 152tl, 153 (all), 188–89 (all), 212–13, 220l, 226b, 227tl, 228–29 (all), 230tl, 231cl and b, 267, 298t

Acknowledgements

I would like to thank my publishers, and particularly Nicola Bailey for the wonderful way in which she has put this book together, and for her patience during the long process of selecting the photographs to tell the stories of the gardens. Great thanks, too, to Rosanna Lewis for her careful editing, and Hugh Merrell for his great enthusiasm for the project. To my own team: everyone in the office, who has borne with me, and spared me time while I worked on the book; and to Jennie Spears, who works tirelessly to promote the work I do. I'm enormously grateful to Kristy Ramage, who took down all my words and helped me write the book, and kept track of all the images along the way. My thanks to Rosie Atkins for writing the foreword, and for supporting my work from the very beginning; to Jane Hyslop for her beautiful illustrations for the binding and endpapers of the book; and to William Collinson, my dear partner, for the amazing record he has created of my work. I owe the greatest gratitude to my clients, who have supported my work, and generously allowed me to include it in this book.

Index

Page numbers in *italic* refer to illustrations.

ARNE MAYNARD is a celebrated garden designer known for his large country gardens in Great Britain, Europe and the United States. He draws inspiration from a wide range of disciplines, from architecture to garden history and from interior design to traditional crafts and techniques. Most importantly, he is a passionate and experimental gardener. He is the author of two other books, *Gardens with Atmosphere* (2001) and *Garden Design Details* (2004).

WILLIAM COLLINSON has been photographing Arne Maynard's gardens for more than two decades. His in-depth understanding of Maynard's work allows him to capture perfectly the essence and detail of each garden.

ROSIE ATKINS is the founder and former Editor of *Gardens Illustrated* magazine, and was subsequently Curator of the Chelsea Physic Garden in London.

Published by Merrell Publishers,
London and New York

Merrell Publishers Limited
70 Cowcross Street
London EC1M 6EJ

merrellpublishers.com

First published 2015
Compact hardback edition first published 2024

British Library Cataloguing in Publication data.
A catalogue record for this book is available from the British Library.

ISBN 978-1-8589-4716-7

Produced by Merrell Publishers Limited
Designed by Nicola Bailey
Edited by Rosanna Lewis
Proofread by Marion Moisy

All plant names have been checked against the Royal Horticultural Society's Plant Database, available at rhs.org.uk.

Printed and bound in China

Front cover, frontispiece and overleaf: Haddon Hall, Derbyshire; see page 170

Back cover: La Frateria, Italy; see page 277